Communication
Technology

Series in Communication Technology and Society
Everett M. Rogers and *Frederick Williams*, EDITORS

Communication Technology

THE NEW MEDIA IN SOCIETY

Everett M. Rogers

THE FREE PRESS
A Division of Macmillan, Inc.
NEW YORK

Collier Macmillan Publishers
LONDON

The Free Press
A Division of Macmillan, Inc.
866 Third Avenue, New York, N. Y. 10022

Collier Macmillan Canada, Inc.

Printed in the United States of America

printing number

1   2   3   4   5   6   7   8   9   10

Library of Congress Cataloging-in-Publication Data

Rogers, Everett M.
    Communication technology.

    (Series in communication technology and society)
    Bibliography: p.
    Includes Index.
    1. Telecommunication.   2. Interactive computer systems.
3. Interactive video.   4. Teleconferencing.
5. Videotex systems.   I. Title.
TK5102.5.R55   1986   621.38   85-27555
ISBN 0-02-927110-X
ISBN 0-02-927120-7 (pbk.)

Contents

Note: Asterisked headings in this Contents are case illustrations.

Preface

When my former colleague Sheizaf Rafaeli was a little boy, he asked his parents how the telephone worked. He was told it was like a dog with its tail in Los Angeles and its head in New York. When you pulled its tail in L.A., it barked in New York. Radio worked the same way, except there was no dog. Perhaps that's about all the average user needs to know about computers, satellites, and the other new communication technologies. As many computer users point out, you don't need to know how the motor works in order to drive a car. One may not need to know about RAMs and ROMs and other technical aspects of computer functioning, but the competent communication student of today must know what computers can do and how they function, and, more broadly, how the new communication systems are adopted and with what social impacts.

The purpose of this book is to define the scholarly field of communication technology for readers who want a nontechnical introduction. In the 1980s, an intellectual revolution in communication science began because of the new communication technologies that were changing the nature of human communication in certain very fundamental ways. New courses are being launched in schools of communication, but few appropriate textbooks for these new endeavors have yet to appear. In order to help provide such materials, my colleague Fred Williams and I established a new series of textbooks entitled Communication Technology and Society. The present book is the first in the series.

I wish to acknowledge the two schools of communication that

served as supportive environments during the period of writing this book. At Stanford University's Institute for Communication Research, I learned a great deal from my former colleague Bill Paisley, with whom I co-taught the first course on interactive communication technologies in 1983. Steve Chaffee helped me with the present Chapter 3 on the rise of communication science. Three other associates of my Stanford years—Ron Rice, now at the University of Southern California; Donald Case, presently at UCLA; and Milton Chen, now at Harvard University—collaborated in research studies of the new media, for example, the Green Thumb project in Kentucky. My greatest intellectual debt at Stanford University is to Sheizaf Rafaeli, now a professor at Hebrew University of Jerusalem; his thoughtful theoretical work on the meaning of interactivity helped me greatly in this book.

The other institution to which I owe a great deal is the Annenberg School of Communications at the University of Southern California. It is unique in being a communication school founded to study the new communication technologies, and it has been fortunate to attract faculty and students with a primary interest in this topic. I learned a great deal about the new media while a visiting professor at USC in 1984, and then joined the faculty of the Annenberg School in 1985. My special debt is to my colleague Ron Rice, whose 1984 book, *The New Media*, is an encyclopedic synthesis of the research evidence for many of the main points of the present volume, thus saving me the need to detail all of these studies here. I profited from my discussions with Ron about his research, and from his comments on the present book. Another scholar who influenced my thinking about the media is Fred Williams, founding dean of the Annenberg School in 1973, my co-editor in this series on Communication Technology and Society, and now director of the Center for Study of Communication Technology and Society at the University of Texas at Austin.

Finally, I thank Elizabeth Lopez and Lornalee Spence at the Annenberg School of Communications for straightening out my prose and punctuation with their Xerox 860 word-processor. My Apple IIe and WordStar were useful companions in this writing exercise.

Describing a revolution while it is happening is a tough task. I hope this book proves to be as useful to you as it was difficult for me to write.

Everett M. Rogers
Los Angeles, California

Communication
Technology

CHAPTER 1

The Changing Nature of Human Communication

"Technological change has placed communication in the front lines of a social revolution."

William Paisley, 1985

The word *technology* comes from the Latin root *texere,* to weave or to construct. So technology should not be limited just to the use of machines, although this narrower meaning is often implied in everyday speech. *Technology* is a design for instrumental action that reduces the uncertainty in the cause-effect relationships involved in achieving a desired outcome (Rogers, 1983, p. 12). A technology usually has both a hardware aspect (consisting of material or physical objects) and a software aspect (consisting of the information base for the hardware). For instance, we distinguish between computer hardware (consisting of semiconductors, electrical connections, and the metal frame to protect these electrical components) and computer software (consisting of the coded instructions that enable us to use this tool). Both the software and hardware are essential for any practical use of the computer, but because the hardware technology is more visible to the casual observer, we often think of technology mainly in hardware terms. It is an oversimplification to think of technology as an autonomous, isolated force that is disconnected from the rest of society (Slack, 1984). In this book, we stress the context of the new technologies of study.

One kind of technology—communication technology—is especially important in modern societies such as the United States. *Communication technology* is the hardware equipment, organizational structures, and social values by which individuals collect, process, and exchange information with other individuals. Certain communication technologies go back to the beginnings of human history, such as the invention of spoken language and such written forms as the pictographs on the walls of caves. Mass media technologies (with at least the potential for reaching a mass audience) date from the clay tablets of such early civilizations as the Sumerians and Egyptians. But technologies such as Gutenberg's movable-type printing press did not actually reach a mass audience until the 1830s, with the advent of the "penny press" in the United States. In the decades that followed, such electronic media technologies as film, radio, and television became important. These mass media technologies are mainly unidirectional, allowing one or a few individuals to convey a message to an audience of many. During the 1980s, a different kind of communication technology became important, and it facilitated the exchange of information on a many-to-many basis through computer-based communication systems. Whether you call it "the new communication technologies," "the new media," or "interactive communication," it is obvious that a very basic change is occurring in human communication.

All communication technology extends the human senses of touching, smelling, tasting, and (especially) hearing and seeing. Such extensions allow an individual to reach out in space and time, and thus obtain information that would not otherwise be available (McLuhan, 1965). Media technologies provide us with "a window to the world," and as a result we know more about distant events than we could ever experience directly.

Nature of the New Communication Technologies

The key technology underlying all the other new communication technologies is electronics. Electronics technology these days allows us to build virtually any kind of communication device that one might wish, at a price (Pool, 1983a, p. 6). One special characteristic of the 1980s is the increased number and variety of new communication technologies that are becoming available. Further, and more important, is the nature of how these new media function; most are for many-to-many information exchanges. Their interactive nature is made possible by a computer element that is contained in these new technologies. In fact, what marks the new communication technologies of the post-1980s era as special is not just the availability of such single new technologies as microcomputers and satellites, but the combining of these elements in entirely new types of communication systems—for example, the use of satellites to deliver a wide variety of programming to cable television systems. Certain cable TV systems, such as Qube in Columbus, Ohio, are interactive (allowing household users to send, as well as receive, messages) because they utilize a computer at the head-end of the cable system.

Communication technology has had a very strong impact on the nature of scholarly research on human communication. The issues studied by communication scientists over the past forty years have been affected by the changing nature of communication (as we will show in Chapter 3). In the past, the basic division of the scholarly field of communication has been a dichotomy on the basis of channel: *interpersonal channels*, which involve a face-to-face exchange between two or more individuals, versus *mass media channels*, all those means of transmitting messages such as radio, television, newspapers, and so on, which enable a source of one or a few individuals to reach an audience of many. This classification is mainly on the basis of the size of the audience, with interpersonal channels reaching from one individual up to a small group of fifteen to twenty. Now, scholars (Dominick, 1983, p. 14) recognize a third category, "machine-assisted interpersonal communication," that has certain qualities of both mass media and interpersonal channels yet is different in several important ways from either one (Chapter 2). An example of such machine-assisted interpersonal communication is the telephone; it does not fit into either category of mass media or interpersonal channels because it is neither face-to-face nor one-to-many. Examples of newer communication technologies are: teleconferencing networks, electronic messaging systems, computer bulletin boards, and interactive cable television.

The new interactive technologies have been available only for several years, and they have not yet become very widely adopted in the United States. Their potential impact, however, is quite high. By 1985, about half of American households had cable television, although only a few cable systems were interactive. Less than 1 percent of American households have videotext or teletext. Over the past decade, 20 percent of households accepted video cassette recorders (VCRs); around 15 percent have at least one microcomputer, and in 1985 about 25 percent of the U.S. work force used computers as their primary work tool. From 1980 to 1985, about 95 percent of American elementary and high schools adopted computers, although less than 10 percent of the students were enrolled in a class in which microcomputers were used. So the interactive communication technologies are off to a fast start. But just a start.

What is different about human communication as a result of the new technologies?

1. All of the new communication systems have at least a certain degree of interactivity, something like a two-person, face-to-face conversation. *Interactivity* is the capability of new communication systems (usually containing a computer as one component) to "talk back" to the user, almost like an individual participating in a conversation. The new media are interactive in a way that the older, one-to-many mass media could not be; the new media can potentially reach many more individuals than if they were just face-to-face, although their interactivity makes them more like interpersonal interaction. So the new media combine certain features of both mass media and interpersonal channels.

Interactivity is an inherent property of the communication process, not just of the communication technology itself, and is thus a unique communication concept (Rafaeli, 1984 and 1985). The exact degree to which computer-based communication can approach human interaction is an important question. One measurement of the ability of computers to think is the Turing test, in which a computer's intelligence is measured by its performance in responding to conversational questions in comparison to human performance in the same tasks. Obviously, not all computer communication is interactive; in fact, not all human face-to-face communication behavior is interactive if interactivity means a two-way exchange of utterances in which the third remark in a series is influenced by the bearing of the second on the first. Sheizaf Rafaeli (1984) poses this interesting illustration of a three-message exchange: (1) a sign on a candy machine catches an individual's attention; (2) the individual inserts 35 cents in the machine; (3) the machine dispenses a candy bar. Are candy machines interactive communication media? No, because they are not "intelligent." The third response is not predicated on the bearing of the second exchange on the first. Here we see that not all two-way exchanges are necessarily interactive; automatic, mechanical reaction is not the same as mutual responsiveness. Human response implies listening, attentiveness, and intelligence in responding to a previous message exchange.

Interactivity is a desired quality of communication systems because such communication behavior is expected to be more accurate, more effective, and more satisfying to the participants in a communication process. These advantages usually come at the cost of more communication message exchanges and the greater time and effort required for the communication process (Rafaeli, 1984).

So the most distinctive single quality of the new media is their interactivity, indicating their basic change in the directionality of communication from the one-way, one-to-many flow of the print and electronic mass media of the past century. In interactive communication systems, the individual is active rather than completely passive or reactive.

2. The new media are also *de-massified,* to the degree that a special message can be exchanged with each individual in a large audience. Such individualization likens the new media to face-to-face interpersonal communication, except that they are not face-to-face. The high degree of de-massification of the new communication technologies means that they are, in this respect at least, the opposite of mass media. De-massification means that the control of mass communication systems usually moves from the message producer to the media consumer.

3. The new communication technologies are also *asynchronous,* meaning they have the capability for sending or receiving a message at a time convenient for an individual. For example, say that an electronic message is sent to you on a computer teleconferencing network; you may receive it on your home or office computer whenever you log-on. Unlike a telephone call, electronic messaging systems avoid the problem of "telephone tag," which occurs when you call someone who is unavailable, then when they return your call you are unavailable, etc. Only about 20 percent of business calls directly reach the individual being telephoned. In new communication systems, the participants do not need to be in communication at the same time. The asynchronicity of computer-based communication means that individuals can work at home on a computer network and thus make their workday more flexible. The new media often have the ability to overcome time as a variable affecting the communication process.

I have a friend who likes to watch the "CBS Evening News with Dan Rather," but he seldom gets home from work in time to see the

broadcast. My friend is one of the 20 percent of American households who owned a video cassette recorder by the mid-1980s. So
whenever he arrives on his doorstep, Dan Rather and the CBS evening News is waiting for him. This time-shifting ability of many of the
new communication technologies is one aspect of asynchronicity. In
addition to video cassette recorders, computer-based communication systems and several of the other new media have this time-
shifting capacity.

Asynchronicity is part of the shift of control from the source to
the receiver in a communication system; in this case the control of
time is put in the hands of the receiving individual. With increasing
frequency, this person can determine the most convenient time to
receive a message. Automated teller machines (ATMs) allow one to
bank in an asynchronous way; instead of being a slave to my banker's
hours, I can now do my banking twenty-four hours a day. Such
added convenience is an important reason for the widespread adoption of ATMs by the American public. Telephone-answering machines also provide this time-shifting and/or time-expanding ability
to many.

There are other differences between the new communication
technologies and their older counterparts of radio, television, and
film; many of the differences stem indirectly from such fundamental
distinctions as the interactivity, asynchronicity, and de-massification of the new media. The new media represent an expanded ac-
cessibility for individuals in the audience, with a wider range of
alternative conduits by which information is transmitted and processed. Further, the format or the manner of display of information
is changing (Compaine, 1981). Finally, compared to the one-way
media, the contents of new communication technologies are more
likely to be informational, rather than just entertainment.

Implications for Communication Research

The new communication technologies have elevated the field of
communication research to a high level of importance in human
society. Public and private policy issues swirl around the results of
research being conducted on the new technologies: international
competition and trade conflicts in high-technology; the transition
from an Industrial Society to an Information Society; and growing
concern with socioeconomic and gender inequalities, unemploy-

ment, and other social problems that result from the impacts of the new communication technologies.

Each of the three main characteristics of the new communication technologies has implications for the conduct of communication research (as we detail in Chapter 7):

1. The interactivity of the new media is made possible by computers, which provide new data and allow use of different data-gathering and analysis methodologies than in the past. The computer element in the new communications systems can retain a complete, word-for-word record of all communication messages in its memory. This record is available for analysis by communication scholars, who in the past have seldom had access to such a gold mine of data about human interaction.

2. The individualized, de-massified nature of the new media makes it almost impossible to investigate a new communication system's effects using the linear-effects paradigm followed in much past research on mass media communication, where a relatively standardized content of the media could be assumed (at least to the extent that the same messages were available to everyone in the audience). With the new media, message content becomes a variable; each individual may receive quite different information from an interactive communication system.

Consider the some 3,000 scholarly research publications on the effects of television violence on children. This inquiry has followed generally a linear, one-way model of communication, exploring whether a consequence of the violent content of children's television programming is aggressive behavior by youthful viewers. Most American children are exposed to the same dose of violent television content. Could this research approach be used to study the effects of the highly individualized content of a computer bulletin board? No. Conventional research methodologies and the traditional models of human communication are inadequate. That's why the new communication technologies represent a new ball game for communication research.

3. The asynchronous nature of the new communication systems also implies major changes in communication research and theory. This lack of time-boundedness makes such machine-assisted interpersonal communication more similar to certain mass communication (you can read today's newspaper today or tomorrow) than is face-to-face interpersonal communication, although its two-person nature is similar to interpersonal exchange. Such asyn-

chronous communication forces researchers to give more attention
to time as a variable than they have in the past when the over-time,
process nature of communication was almost entirely ignored,
perhaps because past communication research methods are suited
best to gathering one-shot data and analyzing it with cross-sectional
statistical methods.

A technological determinist (someone who feels that technology
is the main cause of social changes in society) might attribute the
fundamental changes beginning to take place in human communication as being entirely due to the new information technologies,
particularly computers. Many changes can indeed be traced to the
new technologies, but the way in which individuals *use* the technologies is driving the Information Revolution now occurring in the
United States (and in most other Western nations as well as Japan).
Thus, this book takes a human behavioral approach to understanding the nature of communication technologies, focusing especially
on two overriding issues:

1. *Adoption.* Here, the main research questions include: Who
adopts (purchases) a new communication technology (as compared
to who does not)? Why do they adopt? What is the rate of adoption
of a new technology? What will it likely be in the future? How could
the rate of adoption be speeded up or slowed down? Are individuals
(or households or organizations) who adopt one new communication
technology also likely to adopt other new communication technologies? Is there a key communication technology (for example, the
home computer) that triggers the adoption of other communication
technologies?

2. *Social Impacts.* Here the main research questions include:
What are the direct, intended, and recognized effects or consequences of a new communication technology? What are the indirect, unintended, and unrecognized effects or consequences? How
do the new communication technologies affect the older technologies of communication (for example, how will the telephone be
changed by its increasing use for transmitting computer data)? Do
the new communication technologies widen the gaps between the
information-rich (who are usually the first to adopt) and the
information-poor (who adopt later, if at all)?

These two broad research issues are discussed in Chapters 4 and
5, respectively.

The new communication technologies occur in a sociocultural context; other factors (such as governmental policies) accompany the technology. So, is it the new communication technologies, or the variables accompanying them, that cause the social impacts? It is difficult to separate the social impacts of the new technologies from those of their context. Therefore, we consider both communication technologies and their context as explanations of social change. In this book, we are soft technological determinists, viewing technology along with other factors as the causes of change.

After reviewing the history of communication research, my former Stanford University colleague William Paisley concluded: "Technological change has placed communication on the front lines of a social revolution" (Paisley, 1985, p. 34). I agree. How adequately are communication scholars prepared for this new leadership role? Not very, I think. The new technologies demand an epistemological change in communication research, a paradigm shift (as we argue in Chapter 6).

Linear models of communication, based on source-message-channel-receiver components (Shannon and Weaver, 1949), may have been fairly appropriate for investigating the effects of one-way mass media communication (Chapter 3). And, in fact, such effects-oriented research has been the main preoccupation of mass communication scholars for the past forty years or so. But the interactivity of the new communication technologies forces us to follow a model of communication as convergence, the mutual process of information exchange between two or more participants in a communication system (Rogers and Kincaid, 1981). Such convergence communication behavior implies that it it impossible to think of a "source" and "receiver" in a communication system with a high degree of interactivity. Instead, each individual is a "participant."

The distinctive aspects of the new information technologies are forcing basic changes in communication models and in research methodologies (Rice and Associates, 1984). More broadly, the new communication technologies, through their impacts in helping to create Information Societies, are leading to a new set of communication research issues that are beginning to be addressed by scholars. Thus, the Information Revolution is causing a scientific revolution in communication research.

Welcome to the Information Society

In recent years, the United States and several other highly advanced nations have passed through an important transition in the makeup of their work force, the basis of their economy, and in the very nature of their society. Information has become the vital element in the new society that has emerged, and so these nations are called Information Societies.

An *Information Society* is a nation in which a majority of the labor force is composed of information workers, and in which information is the most important element. Thus, the Information Society represents a sharp change from the Industrial Society in which the majority of the work force was employed in manufacturing occupations, such as auto assembly and steel production, and the key element was energy. In contrast, *information workers* are individuals whose main activity is producing, processing, or distributing information, and producing information technology. Typical information worker occupations are teachers, scientists, newspaper reporters, computer programmers, consultants, secretaries, and managers. These individuals write, teach, sell advice, give orders, and otherwise deal in information. Their main activity is not to raise food, to put together nuts and bolts, or to deal with physical objects.

Information is patterned matter-energy that affects the probabilities available to an individual making a decision. Information lacks a physical existence of its own; it can only be expressed in a material form (such as ink on paper) or in an energy form (such as electrical impulses). Information can often be substituted for other resources, such as money and/or energy—for example, "smart" household appliances that contain a microprocessor save expensive electrical power. Information behaves somewhat oddly as an economic resource in the sense that one can sell it (or give it away) and still have it. Because information is such an abstract phenomena, it is often difficult to perceive its crucial importance in modern society.

CHANGES IN THE LABOR FORCE

Applications of the steam engine to manufacturing and transportation, beginning around 1750 in England, set off the Industrial Revolution that began the transition from an Agricultural Society to an Industrial Society. The Agricultural Society had been the dominant form for about 10,000 years up until this point (and most Third World nations are still Agricultural Societies today). The Industrial Revolution spread throughout most of Europe, to North America, and later to Japan. Figure 1–1 shows that the United States began to industrialize in the mid-1800s; from 1900 to 1955, the largest part of the American work force was employed in industrial jobs. Then, in

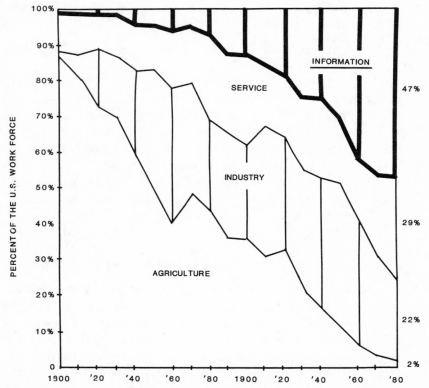

Figure 1–1. The United States became an Industrial Society around 1900 and an Information Society around 1950

The Industrial Revolution, caused by the applications of the steam engine to industrial manufacturing and transportation, began in England about 1750 and spread to the U.S., where the Agricultural Society gave way to the Industrial Society around 1900. Then, in 1955, the Communication Revolution, centering around applications of the computer, began to occur and information workers displaced industrial workers as most numerous in the U.S.

Source: James R. Beniger, *The Control Revolution* (Cambridge, Mass.: Harvard University Press, 1986). Used by permission of Harvard University Press.

1955, a historical discontinuity happened when industrial employment began to decrease and information workers became more numerous. Today they are in the majority. What occurred was a *Communication Revolution*, the social changes in society resulting from the impacts of communication technologies, especially the computer. While the United States led other nations in becoming an Information Society, Canada, England, Sweden, France, and other European nations are not far behind.

The best estimates (Strassmann, 1985, p. 56) available in the mid-1980s indicated that:

- 54 percent of the American *work force* are information workers.
- 63 percent of all *equivalent working days* in the U.S. are devoted to information work (the difference of 9 percent over the 54 percent of information workers is because about one-quarter of the time of all noninformation workers is devoted to information work, while almost none of information workers' time is involved in handling goods or materials).
- 67 percent of all *labor costs* in the U.S. are for information work, as information workers receive wages and benefits that are 35 percent higher than noninformation workers.
- 70 percent of *work hours* in the U.S. are devoted to information work, as information workers put in an average of 10 to 20 percent more work hours per week than do other occupations.

By any of these measures, the United States is definitely an Information Society. Will the recent trends toward information work level off by, say, the year 2000? Probably not. A basic factor in the Communication Revolution is the increasing availability of new information technologies, and many of these communication tools are only in partial use today, so their full impacts have not yet been felt. It is also certainly true that even newer communication technologies are yet to come.

FROM MASSIFICATION TO INDIVIDUALIZATION

Industrial Society was a mass society: mass production, mass media, mass culture. Standardization of products, assembly-line production, and returns-to-scale were aspects of this massification. An Information Society is a more individualized society, de-massified in nature. The new communication technologies make this so. "Such devices as teletext, viewdata, cassettes, cables, and videodiscs all fit the same emerging pattern: They provide opportunities for individuals to step out of the mass homogenized audiences of newspapers, radio, and television and take a more active role in the process by which knowledge and entertainment are transmitted through society" (Smith, 1980, p. 22). Such de-massification of mass media communication represents a shift in control, from the producer to the consumer.

This basic change is a fundamental aspect of the Communication Revolution, leading from the Industrial Society to the Information Society. Certain of the other important characteristics of these two types of societies, and of the Agricultural Society, are compared in Table 1–1.

WHY *Information?* WHY *Now?*

The transition of the United States to an Information Society has been the focus of considerable scholarly research.[1] Such research questions have been pursued as (1) how best to index the progress of a

Table 1–1. Comparison of the Agricultural Society, Industrial Society, and the Information Society

Key Characteristics	Agricultural Society	Industrial Society	Information Society
1. Time period	10,000 years (and continues today in most Third World countries)	200 years (began about 1750 in England)	? years (began about 1955 in the U.S.)
2. Key element/ basic resource	Food	Energy	Information
3. Main type of employment	Farmers	Factory workers	Information workers
4. Key social institutions	Farm	Steel factory	Research university
5. Basic technology	Manual labor	Steam engine	Computer and electronics
6. Nature of mass communication	One-way print media	One-way electronic media (radio, film, television)	Interactive media that are de-massified in nature

nation toward becoming an Information Society (the percentage of
the work force that are information workers is most frequently
utilized), and (2) what social problems (for example, unemployment, inequality, and information overload) typically accompany
the change to an Information Society. Unfortunately, little scholarly
attention has been devoted to such fundamental issues concerning
the Information Society as "Why *information*?" and "Why *now*?"

One theory to explain such "why" questions is proposed by James
Beniger (in press), whose analysis suggests that the Information
Society emerging in the United States since the mid-1950s results
from social changes begun a century ago. In the 1850s, steam energy
technology was applied to manufacturing and transportation,
leading to the Industrial Revolution. The processing of material was
greatly speeded up; for example, the newly constructed railroads
made it possible to move people and products around the nation
relatively rapidly and at low cost. But the Industrial Revolution also
led to a "crisis of control" around 1900 as the ability to control the
new energy technologies lagged behind their widespread use. For instance, Beniger documents the problem of "lost" railroad cars
during this period; effective technologies for keeping track of the
rolling stock did not keep up with the ability of the railway lines to
physically move their cars around the country. The crisis of control
created a need in America to exploit information activities. The
technological means for doing so arrived in the post–World War II
era with the computer and other communication technologies. So in
recent years, according to Professor Beniger's theory, we have both
the need for information-handling activities (stemming from the Industrial Revolution) and the technological tools to meet this need.
The result is the Communication Revolution, and today's Information Society.

Certainly the technical advances in microelectronics that occurred in the 1970s and 1980s have spurred the Communication Revolution. But government policies favorable to the new communication
technologies have also aided their rapid diffusion and adoption in recent years. For example, during the 1980s, the Federal Communication Commission (FCC) reversed its previous policies of protecting
such existing mass media as broadcast television, usually at the expense of new technologies. The trend toward deregulatory policies
on the part of the FCC has opened many parts of the communication
industry to free competition. These hands-off policies generally aid
the diffusion of the new communication technologies. Sometimes,
however, deregulation can have the opposite effect, as in the case of

videotext and teletext, whose development has been slowed by the FCC's refusal to select a standard (Singleton, 1983, p. 4). In any event, one cannot leave the role of government policies out of a thorough understanding of the Communication Revolution.

One might wonder if the U.S. will someday become a post–Information Society, whose prerequisites are now being created by the Communication Revolution. What will be the possible nature of this post–Information Society?

THE RESEARCH UNIVERSITY IN THE INFORMATION SOCIETY

Fundamental to the growth of the Information Society is the rise of knowledge industries that produce and distribute information, rather than goods and services (Machlup, 1962). The university produces information as the result of the research it conducts, especially basic research,[2] and produces information-producers (individuals with graduate degrees who are trained to conduct research). This information-producing role is particularly characteristic of the fifty or so research universities in the United States. A *research university* is an institution of higher learning whose main function is to perform research and to provide graduate training.

The research university fulfills a role in the Information Society analogous to that of the factory in the Industrial Society. It is the key institution around which growth occurs, and it determines the direction of that growth. Each of the several major high-technology regions in the United States is centered around a research university: Silicon Valley and Stanford University in California; Route 128 and MIT near Boston; Research Triangle and the three main North Carolina universities (Duke, North Carolina State, and the University of North Carolina).

A *high-technology industry* is one in which the basic technology underlying the industry changes very rapidly. A high-tech industry is characterized by (1) highly educated employees, many of whom are scientists and engineers; (2) a rapid rate of technological innovation; (3) a high ratio of R & D expenditures to sales (typically about 1:10); and (4) a worldwide market for its products. The main high-technology industries today are electronics, aerospace, pharmaceuticals, instrumentation, and biotechnology. Microelectronics, a subindustry of electronics centered on semiconductor chips and their applications (such as in computers), is usually considered the highest of high technology because the underlying technology changes more rapidly than in other high-technology industries.

Microelectronics technology, applied in the form of computers (especially microcomputers) and telecommunications are driving nations such as the United States into becoming an Information Society. That is why the role of the research university is so important in understanding the emergence of the Information Society.

Because information is such a highly valued commodity in an Information Society, those individuals who produce new information (scientists, R & D workers, and engineers) are treated as the super-elites of the Information Society (Bell, 1973). Thus, the new type of society that is emerging has a new class structure. The amassing, possession, and control of capital allowed the Rockefellers, Carnegies, and Morgans to profit from owning oil fields, steel mills, railroads, and other industrial enterprises during the Industrial Revolution. Today, access to the scientific upper class is mainly through formal education (especially at the graduate level), through the use of intelligence, rather than through the control of capital as it was for the robber barons of the Industrial Society. To the scientific-technological elites of the Information Society will come money and political influence, but their stock-in-trade is brainpower.

There is an interesting historical paradox in the contemporary role of the research university in the Information Society. Many of these universities were founded by the wealthy robber barons of the Industrial Society—Rockefeller at the University of Chicago, Leland Stanford at Stanford University, and Mellon at Carnegie-Mellon University. Thus were the economic gains from the Industrial Society of the late 1890s converted into the key institutions of the Information Society.

The MCC Moves to Austin*

One of the most important recent developments in high technology is the 1983 decision of the Microelectronics and Computer Technology Corporation (MCC) to locate in Austin, Texas. Within a few years of this announcement, Austin became a boomtown, with new housing values shooting up 20 percent during 1984 alone.

The trend in recent years is to closer university-industry relationships, especially in the conduct of microelectronics research. During the 1980s, the federal government cut back severely on funding university research in many fields. Consequently, American universities looked to private industry for research funds. The National Science Foundation estimates that industry funding of university research increased fourfold in 1974–1984, to about $300 million. During the 1980s, many state and local governments launched initiatives to encourage the development of high-technology industry in order to create new jobs and fuel economic growth. Fearful of Japanese competition, American microelectronics firms formed university-industry collaborative research centers, and invested considerable resources in funding these university-based centers.

Largest of the new R & D centers is the Microelectronics and Computer Technology Corporation, located on the campus of the University of Texas at Austin. Fifty-six other cities in twenty-seven states competed with Austin for the MCC, with state and local governments offering a variety of incentives. Here is what 300 Texas leaders in state and local governments, universities, and private companies put together over a two-month period to win the MCC:

- Twenty million dollars in single-family mortgage loans below prevailing interest rates; a relocation office to facilitate utility hookups and to help place the spouses of MCC employees in jobs; $500,000 to underwrite company relocation expenses; use of a Lear jet and a two-member crew for two years.
- Twenty acres of land leased for a nominal fee of $1 per year at the University of Texas's Balcones Research Center; construction and lease of a 200,000-square-foot building (cost: $20 million); and interim office space.
- From the University of Texas: $15 million in endowed faculty positions; creation of thirty-two new faculty positions in microelectronics and computer science; $750,000 in graduate fellowships; and an additional $1 million for research in microelectronics and computer science.
- From Texas A&M University: more endowed chairs in microelectronics and computer science; completion of an engineering research building with space for research; adjunct faculty status for MCC research staff.

Bruce Babbitt, governor of Arizona, whose state was a finalist in the selection process, stated: "Some sixty mayors and twenty-seven governors complained about the unfair advantage of Texas oil money, and promised their constituents a better showing next time." Certainly the MCC decision dramatically heightened awareness among state and local officials about the importance of high-technology development, and created a fuller realization of the role of research universities in attracting high-technology firms.

What did the University of Texas, the state of Texas, and city of Austin get in return for their offer to the MCC? Jobs and an infusion of big money to the state. The MCC is supported at $75 million per year by a consortium of twenty U.S. firms that are the giants of the microelectronics industry, plus U.S government research grants (mainly from the Department of Defense)—the MCC has a research staff of about 400. During its first year of operation, the MCC created a boomtown mentality in Austin. Fourteen high-technology firms moved all or part of their operations, employing 6,100 people, to Austin during 1983; compare this to 1982 when only four companies with 900 jobs moved to Austin. The average selling price of a new single-family home rose 20 percent to $106,157 during 1983.

But the main benefits to Austin of getting the MCC will occur years from now, when a high-technology complex of microelectronics firms develops around Austin. It is possible that this complex may eventually rival or surpass California's Silicon Valley as a center for the production of information technology. In this sense, the MCC's decision may have settled the location of the future capital of the Information Society.

The MCC is only one of about twenty-five university-industry micro-electronics research centers at American universities. Other such collaborative R & D centers have been founded for robotics, biotechnology, and other high-technology fields. The intimate relationships of industries and universities represented by these R & D centers of the 1980s are of obvious advantage to both parties. But they also raise some troubling questions about possible conflicts of interest for the faculty involved, and about potential misallocation of research resources when scientific investigation is driven by corporate interests instead of the spirit of free intellectual inquiry.

Governing the Future Information Society

Capitalism and socialism came out of the Industrial Revolution that occurred in Western nations, and both are certain to be transformed in the Information Society. What form of economic/political philosophy best fits the conditions of the Information Society?

Adam Smith saw the fixed relationships of the feudal period breaking down with the rise of capitalism in Europe. Under capitalism, decisions that had been made by a hierarchical order began to be made by the "invisible hand" of free market forces. Each individual and private firm pursued its own best interests, oriented to maximizing its profits. Such private greed, when aggregated to the level of an industry, a city, or a nation, seemed to result (through the invisible hand) in maximizing public good. Adam Smith argued that when government intervened with a system of free competition, such as by paying a subsidy to a firm or by regulating an industry, inefficiency usually resulted. Thus, a brand of capitalism emerged, especially in the U.S., which believed that the best government was one that governed least.

Karl Marx and the other founders of socialism looked at the Industrial Revolution and saw a very imperfect type of society: wide socioeconomic inequalities, exploitation by capitalists of laborers, and alienation of these workers. Socialism assumed that the route to a better future lay through a proletariat revolution in which the laboring class gained control of society, and then founded a government that would manage the nation toward a more perfect state. So in contrast to capitalism's faith in market forces for making decisions in society, socialism believes in a much greater degree of government intervention in society.

An Information Society changes rapidly, often creating social problems. For example, very major adjustments must be made by the labor force to the unemployment situation caused by the new communication technologies. Millions of factory workers from the Industrial Era are suddenly unemployed, at the same time that many types of information work are created, with most of the unemployed lacking the formal education or job skills for the new positions. How is society to cope with these sudden social changes? What is the role of government, of private companies? Is capitalism, or socialism, or some other economic/political form that is yet to emerge, most appropriate for governing the Information Society?

A crucial question about the Information Society is whether our lives will be better than in the Industrial Society. Undoubtedly there will be "winners" and "losers." Probably there will be a less-equal distribution of socioeconomic resources. The Information Society will likely bring with it a new set of social problems, such as the massive adjustments of the work force necessitated by the change from industrial work to information work.

There are two opposing views of the new society (Mariem, 1984):

1. *An uncritical euphoric stance:* This view is widespread today and is promoted especially by commercial interests. Seldom do these descriptions of the new marvels explore the possible negative consequences of the new communication technologies, or of the Information Society they are helping to create.

2. *A hypercritical, pessimistic stance:* Those who hold this view perceive all the new technologies as leading to disaster. Some of these critical observers even call for a moratorium on the production and adoption of new communication technologies. The pessimists

seldom offer constructive guidance for shaping positively the Information Society.

We try to avoid either of these extreme views in this book. Undoubtedly the new media are being oversold, which is dangerous. Undoubtedly they bring with them important social problems, which we should understand and prevent, or at least mitigate.

A Kentucky Farmer Joins the Information Society*

In 1981, while I was interviewing a Kentucky farmer as part of an evaluation study of the Green Thumb Box (a videotext system providing market, weather, and technological information to farmers), a postal employee delivered the daily mail to the farmer's door. On that particular day, my farmer-respondent received about thirty-five pieces of mail: a local newspaper, the *Louisville Courier,* and the *Wall Street Journal;* thirteen magazines (he subscribed to ten of these and three were free); a dozen or so first-class letters (this farmer was a seed-grower, and several of the letters were related to his business); several bills; a research report that he had requested; and several pieces of junk mail. That day's mail was a foot-high stack that would not fit in the farmer's mailbox, which was why his postman brought it to the door.

This farmer told me that he spent, on average, approximately three hours each evening reading through his mail. He felt this information work was the most important, profit-making task of his farming role. My respondent said that his grandfather had believed that hard physical work was the key to successful farming. His father had believed that close attention to the marketing of his farm products was fundamental to success as a farm businessman. My respondent argued that today's agriculture exists in the context of an Information Society, and so the gathering, processing, and outputting of information is one of the most important roles for a modern farmer. In fact, when I asked the farmer about his use of the Green Thumb Box, he expressed an unmet need for satellite weather maps of the Ukraine (this farmer also grew wheat, and bought and sold wheat futures on the Chicago Board of Trade).

Perhaps in four or five years this Kentucky farmer will receive much of his daily print mail in electronic form via a computerized information system.

Summary

Three basic types of communication channels are distinquished in Table 1–2.

Table 1–2. Main Characteristics of Interpersonal, Interactive, and Mass Media Communication Channels

Characteristics of Communication Channels	Face-to-Face Interpersonal Communication	Interactive (Machine-Assisted Interpersonal) Communication	Mass Media
1. Message flow	One-to-few	Many-to-many	One-to-many
2. Source knowledge of the audience	Source has knowledge of the receiver as a single individual	Source may have a great deal of knowledge of the other participants in an interactive system	Source is a media organization with little knowledge of the receivers
3. Segmentation	High (de-massified)	High (de-massified)	Low, the same message is transmitted to everyone (massified)
4. Degree of interactivity	High	High	Low
5. Feedback	Plentiful and immediate	Somewhat limited; may be either immediate or delayed	Highly limited and delayed
6. Asynchronicity (ability to preserve the message)	Low	High for most types of the new media	Low, but high for some media, such as books and newspapers
7. Socio-emotional versus task-related content	High in socio-emotional content	Low in socio-emotional content	Low in socio-emotional content
8. Nonverbal band	Lots of nonverbal communication	Some new media provide nonverbal communication	Visual mass media provide much nonverbal band; audio mass media do not
9. Control of the communication flow	Potential for equal control by the participants	Potential for equal control by the participants	Little control by the receivers of the mass media
10. Privacy afforded	Low	Usually low	High

Notes

1. Examples of this research on the Information Society are: Fritz Machlup, 1962; Daniel Bell, 1973; and Marc Porat, 1978.
2. *Basic research* is defined as original investigations for the advancement of scientific knowledge that do not have the specific objective of applying this knowledge to practical problems. In contrast, *applied research* consists of scientific investigations that are intended to solve practical problems. Applied researchers are the main users of basic research.

CHAPTER 2

What Are the New Communication Technologies?

"If more of the people who write on the impact of new information technologies would read the literature on the impact of written communication, the discussion of the social meanings of computers might still be fragmentary and unsatisfying (how, at this stage, could it be otherwise?) but it would not be so full of vapid grandiosity."

Michael Schudson, 1985

In this chapter we describe each of the interactive communication technologies, showing how these new media evolved out of a historical context of the earlier writing, printing, and electronic media. Technology is an important cause of the social changes happening in a nation. This crucial role of technology is one justification for the present chapter (and for this entire book). *Technological determinism* is the degree to which a technology is the main cause of social changes in a society. We seek to avoid an extreme position of technological determinism regarding the new communication technologies while at the same time acknowledging that the new media, along with other factors, have shaped the Information Society.

The ultimate consequences of a new communication technology seldom are known or can be very accurately predicted when the new medium first becomes available (Pool, 1983a, pp. 5–6). There is too much uncertainty associated with the new technology. But after a period of some years, the new technology and its potential applications gradually become more thoroughly explored. What emerges may be entirely different from what was originally anticipated. An illustration of this is the 1920s motion picture, which was black and white, silent, and shown in public; 1980s video can be color, sound, and seen in the home or elsewhere. Meanwhile, the film industry established film studios, unions, careers, theaters, and other institutional patterns. Important changes have happened in the film/video industry over the sixty-year period, but the institutionalization of the industry has acted as a powerful constraint on technological change. There is an important point here for the new communication technologies of today: They may turn out quite differently tomorrow than what we presently expect.

We are not breathlessly enchanted by the new communication technologies, given to raving about their potentially powerful accomplishment in a "Gee whiz" manner (although plenty of such literature exists). I feel, however, that the interactive communication technologies of the 1980s are markedly different from their predecessors, and that the nature of human communication is thus altered in very profound ways. Otherwise, I would not have written this book.

At the other extreme, I do not think all of the new communication technologies are inherently dangerously negative in their impacts on human society. Many writers and speakers hold this view, and it is not entirely without some justification. Many of us have encountered computer communication in a negative way: incorrect bank statements that are blamed on "the computer"; plane reservations that cannot be made because the computer is "down"; friends who lose their jobs because of office automation or factory robotization. Computers are widely distrusted, feared, and suspected by much of the public. In this book, we do not treat computers and other new communication technology as entirely a negative factor. Rather, we know from scholarly research that the new technologies have both positive and negative impacts on individuals and society. Our hope is to apply communication theory and research in ways to maximize the potential benefits of the new media.

Four Eras in the Evolution of Human Communication

Twenty-seven main events in the evolution of human communication are listed in Table 2–1, most are key inventions of new com-

Table 2-1. A Chronology of Human Communication

35,000 B.C.	Language probably exists in the Cro-Magnon period.
22,000 B.C.	Cave paintings by prehistoric men.

I. *The Writing Era of Communication* (4,000 B.C. to the present)

4,000 B.C.	Sumerians write on clay tablets.
1041 A.D.	In China, Pi Sheng invents movable type for book printing.
1241	Metal type is substituted for clay characters in Korea.

II. *The Printing Era of Communication* (1456 A.D. to the present)

1456	The Gutenberg Bible is printed with movable metal type and a hand press.
1833	Mass-circulation media begin with the first "penny press" newspaper, the *New York Sun*.
1839	A practical method of photography is developed by Daguerre, which is utilized by newspapers.

III. *Telecommunication Era* (1844 A.D. to the present)

1844	Samuel Morse transmits the first telegraph message.
1876	Alexander Graham Bell sends the first telephone message.
1894	Motion pictures are invented and the first films are shown to the public.
1895	Guglielmo Marconi transmits radio messages.
1912	Lee de Forest discovers the amplification qualities of the vacuum tube.
1920	First regularly scheduled radio broadcasting, by KDKA in Pittsburgh.
1933	Television is demonstrated by RCA.
1941	First commercial television is broadcast.

IV. *Interactive Communication Era* (1946 A.D. to the present)

1946	The first mainframe computer, ENIAC, with 18,000 vacuum tubes, is invented at the University of Pennsylvania.
1947	The transistor, a solid-state type of electronic switch that can magnify electronic messages, is invented by William Shockley, John Bardeen, and Walter Brattain at Bell Labs.
1956	Videotape is invented by the Ampex Company in Redwood City, California.
1957	Russia launches the first space satellite, Sputnik.
1969	NASA's first manned space flight to the moon is guided by an on-board minicomputer that measures two feet by one and a half feet, 3,000 times smaller than ENIAC.
1971	Invention of the microprocessor, a computer-control unit (the central processing unit, or CPU) on a semiconductor chip, by Ted Hoff at Intel Corporation, a Silicon Valley microelectronics company.
1975	The first microcomputer, the Altair 8800, is marketed.
1975	HBO (Home Box Office) begins transmitting programming to

continued

Table 2–1. *Continued*

	cable TV systems by satellite, thus setting off the rapid growth of cable TV in the U.S. (reaching 40 percent of households by the mid-1980s).
1976	The first teletext system is provided by two British television networks (BBC and ITV), in which "pages" (frames) of text and graphic information are transmitted to home TV sets (equipped with a special adapter) by normal TV broadcast signals.
1977	Qube, the first interactive cable television system, begins operation in Columbus, Ohio.
1979	The first videotext system is provided by the British Post Office, so that "pages" (frames) of text and graphic information can be sent from a central computer via telephone lines to be displayed on home TV sets.

munication technologies. Notice how the successive interval in years between each has been getting shorter and shorter. The amount of time between a technology's invention and its widespread impact has also been decreasing; for example, about 380 years were required between Gutenberg's invention of an efficient printing technique and the first print *mass* medium (the "penny press," launched by the *New York Sun* in 1833). Not even thirty years were necessary for television, first demonstrated in the 1930s, to reach widespread use in American households at the end of the 1950s. Further, notice the cumulative nature of the communication technologies in Table 2–1; each successive technology did not replace previous communication media (although it usually affected them in important ways). So the history of communication is the story of "more." Each new medium may change the function of previous media, but they do not disappear. Finally, Table 2–1 shows that a great many of the communication inventions of the past 150 years are American. This national dominance stems from the fact that the field of electronics has been a specialty of the U.S. Nevertheless, in recent years some of the inventions and several of the first pilot projects involving videotext, teletext, and "wired cities" have occurred in England, France, Canada, and Japan.

For purposes of simplification, we divide the chronology of human communication into four eras (Table 2–1): writing; printing; telecommunication; and interactive communication.

I. WRITING

For thousands of years, writing was the main method of media communication. Perhaps it can be dated approximately from 4,000 B.C. in Sumeria. Until Gutenberg's invention in 1456 A.D., important books (the Bible, the works of Aristotle or Virgil) were copied by hand by individuals who had a particularly neat handwriting. Much of this copying was done in Catholic monasteries, each of which had a scribe's copying room. Needless to say, until the Renaissance, the amount of reading material available to the average individual was very limited, and the rate of literacy was very low. In medieval libraries, books were chained to the desk where they had to be read. The invention of printing freed books from these chains.

II. PRINTING

Printing was at first called "artificial writing" (as opposed to "natural writing"), in a sense similar to today's "artificial intelligence." Printing technology actually began in China with the invention of making paper from textiles soon after 1,000 A.D. Movable type was also invented in China by Pi Sheng around 1041. This invention was developed for wide use in Korea, with metal type being substituted for clay characters in 1241 (Pool, 1983a, p. 12). Independently, Johann Gensfleisch (whom we know today as "Gutenberg") invented printing with movable metal type in 1456. Gutenberg lived in Mainz, Germany, an area where wine presses were widely utilized, and he modified one of these presses to print words on paper. Also a goldsmith, he understood metal casting and made metal letters that were temporarily placed in frames to print a page, and then reused to print other material. The new technology meant a much wider distribution of books; until Gutenberg, a skilled copyist could reproduce only two books a year. A printer using Gutenberg's press could make one book a *day* (Pool, 1983a, p. 14). Thus, it is not surprising that the date of Gutenberg's invention is usually accepted as the initiation of the 150-year period of the Renaissance (from the French word meaning "rebirth") during which Western civilization broke out of the feudalism of the Dark Ages into the enlightenment of science, university learning, and the rapid expansion of the known world. The widespread availability of books was a key factor in this rebirth of Western civilization.

The impact of printing technology was very gradual, in large part because only a small percentage of the European population were literate. The press did not become a mass medium in a modern sense until 380 years after Gutenberg's invention. On September 3, 1833, Benjamin Day launched the *New York Sun* with the motto "It shines for all." His newspaper was vulgar, sensational, and cheap. The era of the "penny press"—newspapers that sold for one cent and were aimed at a mass audience of common Americans—thus began. Large press runs of 30,000 or 40,000 were made possible by high-speed printing presses. So the first mass media revolution was part of the Industrial Revolution in America: mass production due to power machinery, assembly lines, and the factory system (Pool, 1983a, p. 18).

III. TELECOMMUNICATION

The two root words in telecommunication imply its meaning: communication at a long distance. One important function of telecommunication is to provide a substitute for transportation; instead of moving people to ideas, telecommunication moves ideas to people.

The new telecommunication technologies of the period beginning in the mid-1800s dispensed with the need for literacy, providing a means of jumping this barrier to mass communication. Because radio and television utilized the public airways, the federal government felt justified in regulating the broadcast media through issuing licenses to radio and television stations. The Radio Control Act of 1927 and the Communication Act of 1934 established the FCC (Federal Communication Commission) in order to resolve frequency interference problems caused by the proliferation of new stations.

Several of the most important telecommunication technologies (radio, film, and television, for example) were primarily one-way, one-to-many mass media, but several, such as the telegraph and telephone, were mainly one-to-one and interactive in nature. Actually, a certain degree of feedback from the one-way electronic media occurred, such as through letters-to-the-editor, television ratings, and audience surveys. But compared to the new communication technologies of the 1980s, the electronic media are much less interactive.[1]

The first electronic telecommunication occurred on May 24, 1844, when Samuel Morse, inventor of the telegraph, sent the famous message "What hath God wrought?" from Baltimore to Washington, D.C. Until that time, information could travel only as fast as the messengers who carried it; communication of information and transportation of people and materials were not separated in meaning (Czitrom, 1982, p. 3). But the telegraph changed all that; a network of "lightning lines" soon crossed the nation. The electrical messages that crackled along these wires were many times faster than the fastest trains, whose rails the telegraph wires paralleled.

How the Telegraph Impacted Newspapers*

The telegraph in America provides an early example of how a new communication technology causes powerful impacts on an existing mass medium. In 1844, American newspapers contained little of what is today considered national and international news. "Telegraphy made possible, indeed demanded, systematic cooperative news gathering by the nation's press" (Czitrom, 1982, p. 16).

The Associated Press (AP), a cooperative news agency formed in 1849, originally consisted of six New York newspapers that collaborated in gathering and distributing international news. As telegraph lines spread to more and more U.S. cities, so did the number of newspapers served by the AP. In fact, the gradual monopolization of telegraphy by Western Union grew together with the monopolization of news gathering by the AP; both organizations profited directly from their alliance. Western Union refused to carry the news transmissions of competing news agencies, and the AP utilized only Western Union telegraph services.

The Civil War in the first half of the 1860s (about twenty years after the invention of the telegraph) provided a strong impetus for cooperative news gathering, a process in which telegraphy was the crucial catalyst. Each American newspaper could not afford to have its own correspondents reporting from each of the Civil War battlefields. The Associated Press, utilizing newly installed telegraph services, could uniquely fill this information need. In order to provide acceptable news coverage to the wide variety of political viewpoints represented by American newspapers in the 1860s, the AP had to treat the Civil War news in an "objective" manner, sticking to facts as much as possible and avoiding a subjective, political interpretation of the news. Thus did objectivity become the hallowed professional value of American news-papermen.

The telegraph affected American journalism in yet other ways. For example, AP correspondents were instructed to put the most important facts in the first sentence of each news story, with less crucial information in each succeeding sentence. This "inverted pyramid" style of news writing was necessary because telegraphic services from the Civil War battlefields were occasionally disrupted, and it was essential for the highest-priority information to get through to newspapers using the AP service. The inverted-order style of writing was also convenient for the newspaper editors who used the AP services; they could cut an article to any length to fit available space. Today, every student in an introductory journalism course is taught the inverted pyramid technique of news writing, and the need for objectivity.

Daniel Czitrom, in his history of the mass media, concludes that the telegraph "transformed the newspaper from a personal journal and party organ into primarily a disseminator of news" (Czitrom, 1982, p. 18). That is a rather strong statement of technological determinism, but it does seem to fit the historical facts. And new communication technologies have been impacting the existing mass media ever since the Civil War.

IV. Interactive Communication

The contemporary era of person-to-person telecommunication centers around two-way media that are made possible by computers. Thus it is appropriate that we date the beginning of the interactive communication era from 1946, when ENIAC, the first mainframe, was invented in Philadelphia (Table 2–1). But thirty years of further research and development were needed before miniaturized, low-priced microcomputers became available, a type of computer that many households, businesses, and schools could afford. So by the 1980s, enough individuals had access to computers for them to become an important means of communication.

One might have the following mental image of the mass media audience for radio, television, newspapers, and magazines in the United States: individuals interested in the media messages they receive, eager to obtain information, and active in accessing the media. In this idealized view, the public would be well-informed and alertly up to date in what's going on in the world.

Unfortunately, this image of the U.S. media audience is completely wrong. Audience research by communication scholars shows that most individuals do not really pay much attention to the mass media, they do not learn much from them, and they do not know much about the news of the world. In fact, most people just let the mass media sort of wash over them. Television has become a kind of video wallpaper; a large percentage of Americans passively watch their "least objectional program," absorbing little of the message content. When asked to recall certain salient facts from a TV news broadcast within a few hours of viewing it, few can do so. Many new TV sets and video cassette recorders come with a remote control that makes it easy to switch channels or to "zap" commercials, but only 2 or 3 percent of all viewers do it. So the general picture that emerges of the mass media audience in America is that of passive receivers.

The new communication technologies lead to strikingly different communication behaviors in that they require a high degree of individual involvement. A person must actively chose the information content that he or she wants. Consider someone with a microcomputer and a modem who is accessing an item of desired information (such as an airline schedule) from a data-bank such as The Source or CompuServe, or perhaps this individual is interacting with other members of a computer bulletin board. Clearly the era of the passive media audience is gone, or at least going.

The new electronic technologies are causing an integration of media that we have conventionally considered to be completely separate; for example, the French Minitel system is a device that combines a microcomputer function, a television screen, and a telephone wire in order to provide an electronic telephone directory. The Minitel is a small TV-like screen that attaches to a phone, and connects the user to 100,000 pages of telephone directory listings (including all the telephone numbers in France). A few years ago, when I lived in France, the phone directories for the city of Paris took up four feet of shelf space in my apartment. So the 300,000 Minitels in use in 1985 saved the lives of a great many trees. Three million Minitels are expected to be in use, worldwide, by 1987. (Several pilot projects using Minitels have been carried out in the U.S.)

Today, all of the mass communication media are becoming electronic (including newspapers and book printing), and this convergence of modes is upsetting the three-way division of communication regulatory structures that had evolved in the United States (Pool, 1983a, p. 233):

1. Print, which is generally free of government regulation
2. Broadcasting, in which the government licensed private owners

3. Common carrier (such as the postal system, telephone, and telegraph), in which the government assured nondiscriminatory access for all.

The integration of the communication media caused by the new technologies, such as the way computers and telephones are combined in computer teleconferencing, is paralleled today by an integration of media corporations. For example, American Telephone and Telegraph (AT&T) is now manufacturing and selling computers. Warner Communications (a multinational conglomerate involved in film, music and records, and other media) operates Qube, the interactive cable TV service pioneered in Columbus, Ohio, as a joint venture with Sherson/American Express, the credit card company. Increasingly, a company that one would not ordinarily think of as a communication enterprise is branching out into commercial activities providing a new communication service. This widespread merging and integration of communication media institutions is a direct result of the new communication technologies, which unfreeze the formerly tight boundary around the private mass media institutions (newspapers, TV stations, film companies, radio stations, etc.). This integration of conventional media institutions with nonmedia firms is happening not just in the United States, but also in France, Germany, Japan, and other Information Societies.

What are the main new communication technologies that are included in the term "the new media"? Here is a list:

1. *Microcomputers* are stand-alone units, usually with provision for individual loading of software, and sometimes connected with other microcomputers in a network. The central processing unit (CPU) of a microcomputer, which reads and executes program instructions, is a single semiconductor chip (a microprocessor).

2. *Teleconferencing* is a small group meeting held by interactive electronic communication among three or more people in two or more separate locations. The three main types of teleconferencing are video teleconferencing, audio teleconferencing, and computer teleconferencing.

3. *Teletext* is an interactive information service that allows individuals to request frames of information for viewing on a home television screen; these frames are transmitted in the vertical blanking interval of a conventional television broadcast signal. The lines of information for teletext are located above the picture seen on a television screen. Each of several hundred frames can be chosen by

an individual via a keypad and, after decoding, viewed on a television receiving set.

4. *Videotext* is an interactive information service that allows individuals to request frames of information from a central computer for viewing on a video display screen (usually a home television receiver). The number of frames is potentially unlimited, other than by the capacity of the computer in the videotext system. Videotext requires a request channel (unlike teletext), so it is much more interactive in nature.

5. *Interactive cable television* provides for the sending of text and graphic frames, as well as full video pictures, to home television sets via cable, in answer to a request. The amount of content is potentially unlimited, other than by the head-end computer's capacity. The cable serves also as the request channel. The source computer is usually capable of polling and tabulating responses, accepting orders for services or products, etc. The Qube system in Columbus, Ohio, is an example of an interactive cable system.

6. *Communication satellites* relay telephone messages, television broadcasts, and other messages from one place on the Earth's surface to another. The satellite is usually placed in a stationary orbit around the equator, about 22,300 miles from the Earth's surface. Essentially, satellite transmission of television, telephone, and other information removes the effect of distance on the cost of communication. In the United States, satellites have been utilized since 1975 to distribute television programming to cable TV systems; the rich variety of channel choices thus provided to American households has caused a rapid increase in the rate of adoption of cable TV.

Figure 2-1 shows how these and other communication technologies range on the continuum of interactivity. The traditional mass media of newspapers, radio, film, and television are relatively low in their degree of interactivity. Teletext is at a midway point on the interactivity continuum, and the several varieties of computer communication are relatively high in their degree of interactivity.

Computer Communication[2]

The electronics era really began in 1912 in Palo Alto, California, with the invention of the amplifying qualities of the vacuum tube. Lee de Forest and two engineering colleagues leaned across a table

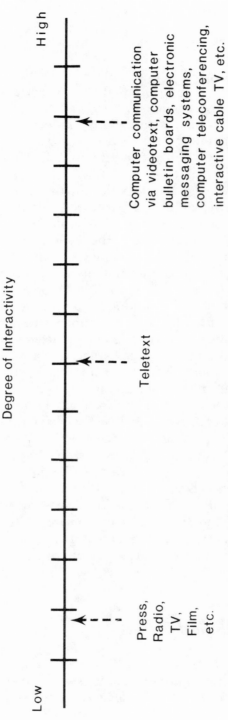

Figure 2–1. Selected Communication Technologies on the Continuum of Interactivity

Interactivity is the capability of new communication systems (usually containing a computer as one component) to "talk back" to the user, almost like an individual participating in a conversation. The mass media of newspapers, radio, television, and film are relatively low in interactivity, although not zero. The degree of interactivity depends not just on a communication technology, as this figure implies, but also on how that technology is used in a particular situation.

and watched a housefly walk across a sheet of paper. They heard the fly's footsteps amplified 120 times, so that the steps sounded like marching boots. This birth of the electronics era opened the way for the later invention of radio broadcasting, television, and computers, all of which were based on electronic amplification.

At first, these electronics technologies utilized vacuum tubes. However, they presented a problem; the tubes burned out, generated a lot of heat, and used considerable electrical power. These shortcomings of vacuum tubes were illustrated by ENIAC, the first mainframe computer. It filled an entire room and used so much electricity that the lights of Philadelphia dimmed when the computer was turned on. In order for computers to become cheaper, smaller, and more widely utilized, an alternative to vacuum tubes would have to be found.

TRANSISTORS AND SEMICONDUCTORS

Bell Labs has the largest basic research program in electronics in the world. Located at Murray Hill, New Jersey, Bell Labs was the R & D arm of the American Telephone and Telegraph Company (until its court-ordered divestiture in 1982). For over fifty years, Bell Labs has spawned a flood of technological innovations; currently, it holds 10,000 patents and produces about one per day. Bell Labs's most significant discovery was the transistor. Some call it the major invention of the twentieth century.

The transistor (short for "transfer resistence") allows the magnification of electronic messages, as do vacuum tubes, but transistors require only a little current, they do not generate much heat, and they are much smaller in size. A transistor is a kind of electrical valve that is used to control the flow of electricity so as to boost, or amplify, the electrical signal. A transistor is a solid-state device, usually made of a semiconductor material such as silicon. Many useful applications of the transistor were expected when it was invented in 1947, but reliable transistors were difficult to manufacture, and the first commercial use did not occur until 1952, five years later, when transistors were used in hearing aids. Gradually, transistor technology advanced, and by the time William Shockley and his two co-inventors received the Novel Prize in 1956, twenty companies were manufacturing transistors. One of these was Shockley Semiconductor Laboratory, in Palo Alto, California.

Shockley started his semiconductor company in Silicon Valley in order to make a million dollars (he said at the time). Unfortunately, his firm was unsuccessful and short-wired. Shockley astutely recruited eight bright young men, but inadvertently taught them the entrepreneurial spirit, and within a year the eight defected to start semiconductor manufacturing companies on their own. Today there are eighty or so semiconductor firms and thousands of other high-technology companies in Silicon Valley (Rogers and Larsen, 1984). None of the firms, nor their founders, would have been in California had it not been for Shockley. So he deserves credit for starting the entrepreneurial chain-reaction, as well as for co-inventing the transistor.

INVENTION OF THE MICROPROCESSOR

Other than the invention of the transistor at Bell Labs, the most significant innovation in the microelectronics industry is the microprocessor, invented in 1971 by Marcian E. (Ted) Hoff, Jr., an employee of Intel, a Silicon Valley semiconductor company. A *microprocessor* is a semiconductor chip that serves as the central processing unit (CPU) controlling a computer. In other words, a microprocessor is the computer's brains. In discovering the microprocessor, Hoff had the inspiration to pack all the CPU functions on a single chip. He attached two memory chips to his microprocessor: one to hold the data, and another to contain the program to drive the CPU. Hoff now had in hand a rudimentary general-purpose computer, a microcomputer.

The basic trend in the semiconductor industry for the past decade or so is toward further and further miniaturization, to put more and more capacity on the same-sized semiconductor chip. This tendency has greatly reduced the cost per bit of computer memory. The most widely sold computer memory chip moved from 1K (1,000 bits of information) in 1973, to the 4K in 1977, the 16K in 1980 the 64K in 1984, and 256K chip in 1986. The cost per bit of computer memory has decreased correspondingly about 28 percent per year for the past dozen years. This lowered cost translates directly into more and more ubiquitous computers in society. Most of the millions of computers sold in the past decade are microcomputers, built around the microprocessor that Ted Hoff invented at Intel.

The Rise of Computer Communication

The idea of communicating with computers has its roots in the fact that mainframe computer power was very expensive. Until the mid-1970s, when the microcomputer started to become popular, computers were owned and operated by the Establishment: government, big corporations, universities, and other large institutions. These mainframe computers were mainly used for such data-crunching tasks as accounting, recordkeeping, research and data analysis, and airline ticketing. Single users could not afford to own a mainframe computer of their own.

The solution to the high cost of large computers entailed creating a computer communication structure in which the mainframe computer was shared. Several users joined a time-sharing system in which a single mainframe was wired so that it would perform more than one task simultaneously, thus "sharing" its time across several users. The technology allowed users to communicate with the large computer over telephone lines. It was soon realized that these telephone connections could be utilized for user-to-user communication. The potential of computers as a special means of communication thus became evident.

Networks emerged, linking users and computers (Figure 2–2). These networks became a kind of "public commons" in which users could communicate with each other and share a common collection of information, such as a data-base. During the 1980s, miniaturization, decreased costs, and the popularization of microcomputers lead to a tremendous explosion in computer networking. Computers had now become a special medium of interactive communication.

At present, about one-quarter of home computer owners have a modem (a computer peripheral that allows one to connect a microcomputer via the telephone system to an information bank—for example, The Source or CompuServe, to another computer, or to a network of other microcomputer users). And in the mid-1980s, about 15 percent of American households had a microcomputer.

Individuals who have access to a microcomputer or a computer terminal can access computer-based data-banks in order to obtain information rapidly and at a relatively low cost. Some data-banks are provided for use by the general public (The Source and CompuServe), while others aim at specialized users. An example of the lat-

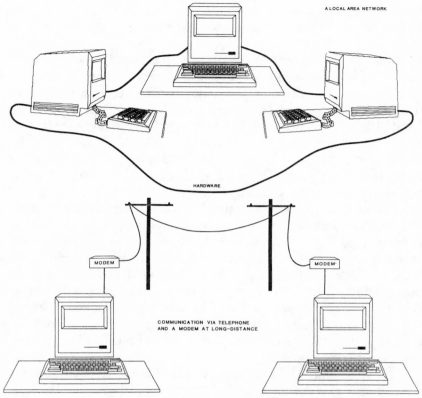

Figure 2–2. Computer Communication May Link Individuals in a Local Area Network (Such as Individuals Working in the Same Organization) or Those Who Are Physically Dispersed (Computer Teleconferencing)

During the 1980s, a tremendous increase occurred in the number of individuals in the U.S. who participated in computer communication networks. When the individuals in a network are located in the same organization, as in an electronic messaging system, their messages can flow through "hard-wiring" that is built into the building. When the individuals in a network area are scattered geographically, their microcomputers are usually linked by a modem and telephone lines.

ter is MEDLARS, which, by certain measures, is the most widely used data-bank in the world. MEDLARS is operated by the National Library of Medicine, a government funded information service located in Bethesda, Maryland, on the outskirts of Washington, D.C. MEDLARS contains a data-base (called MEDLINE) with 4 million medical research reports, and this data-base is continually updated by inputting the articles from 3,000 medical journals. This information bank is searched over a million times per year by doctors and medical staff throughout the United States and abroad. Say a medical doctor practicing in a small town encounters a strange set of symptoms in a patient; she can enter several key words into the microcomputer in her office and be instantly connected, via a long-distance phone line, with all the accumulated medical research literature in MEDLARS. There is a similar computerized data-base for lawyers (called LEXIS), as well as for other specialized professions.

Yet another information bank that can be accessed by a home computer is ISI (Institute for Scientific Information) in Philadelphia. The basic datum that the ISI mainframe computer contains is the citation (a publication that has been cited by the author of a scientific journal article). ISI inputs 7.5 million such citations every year, and its data go back over twenty years. There are many such uses for the ISI data-base; for example, say you want to know who cites a scientific publication that you have written. Or perhaps you might wish to know which scholars are members of an "invisible college" investigating a particular scientific topic. ISI can tell you.

Computer networking is illustrative of the new type of human communication that is the hallmark of the Information Society. Several thousand microcomputer-hosted bulletin boards (the most primitive, but very popular, version of computer networks) were operating in the U.S. in the mid-1980s. In Northern California, for example, a dozen bulletin boards are devoted to political topics, none of which charge its users or produces any profit for its operator. Nevertheless, these bulletin boards are constantly busy.

Computer communication is not just an impoverished attempt to emulate interpersonal interaction. It has several advantages over face-to-face communication, such as asynchronicity, but it also suffers from its lack of a nonverbal band, and has several other differences from face-to-face interaction. Computer designers try to create a computer that can interact with an individual in a way that is similar to human-to-human communication. *User-friendly* is the degree to which a computer and an individual can communicate with the same ease that two individuals can converse. While this definition is fairly obvious, exact measures of the degree of user-friendliness are lacking. We note a certain similarity between our concept of interactivity and the usual meaning of the term "user-friendly." Meanwhile, computer manufacturers each claim their product is user-friendly.

Computers will become the printing presses of the twenty-first century. Publishing is becoming electronic for both reasons of convenience and cost. Large information bases can be edited, stored, transmitted, and searched with a speed and flexibility impossible for ink records on paper (Pool, 1983a, p. 190). Paper will not disappear in the future, any more than typewriters have replaced all pens; paper is too useful for certain purposes.

Nor does computer communication eliminate the need for literacy. In fact, communication via computers demands not only the ability to read, but also the ability to type (as anyone who has faced a computer keyboard certainly knows). The term, "computer literacy," with an inexact meaning, has arisen in the 1980s. For some who use this term, it means the ability to program; for others it means the skill of using available computer programs, while yet others use computer literacy as being prepared to live and work in the Information Society of the future.

COMPUTER BULLETIN BOARDS

One of the best pure examples of an interactive communication system is a computer bulletin board. Each individual participant can communicate directly with any other individual who is connected to the same computer bulletin board. Each of these communities of individuals share a common interest in some topic. Each such computer network has a minimum of social structure that might shape communication flows, so that a computer bulletin board approaches a kind of communication free-for-all. Such electronic networks are very highly interactive in that they closely approach human conversations in nature (except with the important difference that all messages have to be typed into a computer keyboard). Although the percentage of the U.S. population that belongs to at least one computer bulletin board is still very low (probably only 2 to 3 percent), the number of bulletin boards in the U.S. increased tremendously during the 1980s.

A computer bulletin board consists of a host computer that is accessed by participants using their own terminal and/or microcomputer connected to the network by means of telephone lines. A special software package (available commercially) is needed at the host computer in order to route the message traffic. A bulletin board is much like a "local area network" or an electronic messaging system in an organization, except that the bulletin board is a communication system with a completely open membership. Any individual who shares a common interest (for example, antinuclear warfare, computer music, or gay liberation), with other participants on the bulletin board is welcome to join, usually by just dialing a particular telephone number. The participants post notices, make requests for help, and take part in an ongoing discussion of some topic.

My observations of several computer bulletin boards indicate that typically a specific issue is placed on the public "agenda" of the system by some individual, which then sets off a flurry of message exchanges about that topic over the next few days. This discussion may continue for several weeks or so, and then gradually taper off, eventually to be replaced by newer topics. Some typical topics that generated considerable discussion during the fall months of 1984 on some California-based bulletin boards were:

- The decline of Western civilization (launched by a strong statement citing rising divorce rate, drug use, etc.)
- The fairness of the Reagan administration
- Religion and the U.S. Constitution
- The failure of several satellites launched by the U.S. space shuttle.

Scattered among these issues were more humdrum announcements:

- Grader wanted for EE 101
- Scuba club meeting
- How could one secure a microcomputer to a table, so as to prevent theft
- Mercury Monarch for sale

Presumably, a computer bulletin board should provide an extremely democratic communication system; each individual has direct access through the electronic network to every other member of the bulletin board. The communication system is essentially destatused in that there is no social or organizational hierarchy, and gatekeeping (such as by a receptionist or secretary or assistant) is absent because most members of a computer bulletin board process their own electronic mail. Perhaps it is surprising then that, under these conditions, actual participation in bulletin boards is not very equal; a common research finding is that about 50 percent of all messages are sent by only about 10 percent of the participants. That means the other 90 percent of the participants generate only half of all the messages on a system (see Figure 4–3).

In an evaluation of an electronic messaging system at Stanford University, my colleagues and I found that much bypassing of the layers in the organizational hierarchy occurred, especially during the first months of the system. Participants sent many copies of a message to their colleagues, in part because it was so easy to do— "Just push the 'send' button." Such electronic messaging behavior caused complaints of information overload; it also decreased the influence of the organizational hierarchy on communication patterns. For example, in most organizations it is forbidden for an employee to communicate directly with the boss's boss. But when an electronic messaging system is first provided in an organization, there is a tendency to bypass this hierarchy. Participants, especially the executives, soon begin to complain about being overloaded with messages. Gradually, over a period of several months, this problem is resolved as participants in the electronic network learn to be more sparing in sending "cc's" of their messages.

Actually, there is a system of status and prestige on a computer bulletin board, but it may not be based on the usual kinds of organizational status symbols. For example, a computer bulletin board has "sysops," systems operators who are sort of electronic gatekeepers and managers; they are usually computer experts who monitor the flow of messages on the bulletin board, including the right to peek at any message on the system, and even to censor certain content. In essence, sysops are the policemen of a computer bulletin board.

One of the important advantages of a computer bulletin board is its large capacity. An illustration is provided by the Computer Memory Project, a free-access community bulletin board in Berkeley, California, with computer terminals in food co-ops and other public places. For example, use of the Computer Memory Project is replacing thousands and thousands of three by five index cards pinned to the huge bulletin board in the Berkeley Co-op on Telegraph Avenue. The cards described used cars for sale, gripes about local restaurants, and uncensored sex jokes. Now, all these messages go into the computer bulletin board, at the rate of about fifty a day. The main advantage of computerizing the Berkeley bulletin board is its tremendous memory, equivalent to eight football fields covered with the index cards that were previously used.

With that kind of potential capacity, users of the system need a systematic way to search for needed information; that ability is easily provided by the computer. Want to find a thirteen-inch color television set for $100? Just punch in the word *television* on the computer keyboard. . . .

In addition to being highly interactive, asynchronous, and demassified, computer bulletin boards also can be anonymous because most participants use "computer names." In fact, these "handles" show a great deal of sly binary humor; some typical computer names on a Stanford University student bulletin board are:

E.EEYORE
C.CHARISMA
O.OBIWAN
W.WOBEGONE
B.BADWOLF
D.DROID

The anonymous nature of bulletin board communication can lead to a certain degree of honesty and candidness. One bulletin board user commented: "You don't have to wear anything" (Phillips, 1982). This statement could be taken literally or figuratively.

An investigation of LOTS (Low Overhead Time-sharing System), a free student bulletin board at Stanford University, by Rafaeli (1983) showed a high level of use. About one-third of the students and faculty at Stanford, over 5,000 individuals, actively used the LOTS bulletin board. The system carried an average of 300 messages a day, and most users signed on several times a week, spending about fifteen minutes reading accumulated messages each time. Student users of the LOTS bulletin board were especially likely to be engineers, graduate students, and individuals enrolled in computer science classes. Thirty percent of the bulletin board participants in the survey classified themselves as "hackers," someone who is addicted to the use of computers. Many of the messages on the bulletin board were highly utilitarian, dealing with jobs, items for sale, public events, and computing information. A close second in message content was humorous material, such as the use of a popular joke file; a few years ago, Stanford University officials objected to the off-color, ethnic jokes and ordered the file to be erased (it immediately sprang up again).

During the ten years that I taught at Stanford University, I found one of the most effective means of understanding the student body's needs, interests, and opinions was to monitor the LOTS bulletin board. Its content represented a fresher, less-inhibited expression than print alternatives. Unlike the *Stanford Daily*'s letters-to-the-editor, no gatekeeping had occurred through which editors might have neutralized readers' demands for action or blue-penciled out certain expressions. The computer bulletin board also represented a greater volume of student communication activity than did the student newspaper. The bulletin board communication was also more interactive, in that one student's comment would often lead to a response, to which the original writer would reply, etc. Such a high degree of interactivity was unlikely in the letters-to-the-editor column of the student newspaper. So if you want to sample informal public opinion in an organization, check out the system's computer bulletin board.

There is a felt need by most bulletin board users for light content, rather than just heavy doses of information. For instance, even AR-PANET (a bulletin board for defense research contractors provided by the U.S. Department of Defense) carries a large volume of jokes, nonserious messages, and highly personal messages despite the officially staid nature of ARPANET. One communication scholar, Carolyn Marvin, notes that the playful users of ARPANET "form a community united by their achievement in foiling the system," (Marvin, 1983). Perhaps we should not forget that the users of computer bulletin boards, just like everyone else, like to have fun.

Videotext and Teletext

There has been an unbelievable confusion for the past decade[3] about the correct terminology for the types of information services provided to homes in which consumers can request information from a central computer that is transmitted to them via some type of wire, such as a telephone line or cable TV (this service is called videotext), or over-the-air broadcasting (teletext). We define *videotext* as an interactive information service that allows individuals to request frames of information from a central computer via telephone or cable, for viewing on a video display screen (usually a home television receiver). It is essentially a two-way link between a computer and your TV set. *Teletext* is an interactive information service that allows individuals to request frames of information for viewing on a home television screen; the frames are transmitted in the vertical blanking interval of a conventional television broadcast signal (Figure 2–3).

The usual television signal does not use all of the lines of dots allocated to it (525 in the United States, 625 in Japan and most of Europe); those left blank (twenty-one of the 525 in the U.S.) are called the vertical blanking interval and are normally out of sight. These lines can be utilized to transmit additional information, piggybacked on the normal TV signal. Using a keypad that looks something like a hand-held calculator, the viewer selects a page number from a menu or table of contents, the teletext system grabs that frame of information by snatching it out of the flow of electronic data the next time that page is broadcast, decodes it, and then displays it on a television screen. A pause of from ten to twenty-five seconds is necessary for the frame to register on a television screen, and then it can be held there. This wait is needed for all of the flows of information being broadcast in the field blanking interval to pass by a home TV set, so that the requested frame can be grabbed. Typically, teletext systems contain about 400 frames of information; if more are provided, the waiting period will be longer. Given the limited set of information that can be requested, teletext is less interactive than videotext (Figure 2–3). The advantage of teletext is that its cost is low; the consumer only needs to buy a small decoding box to attach to his television set. No monthly fees are involved with teletext because, unlike videotext, it is supported by advertising. Teletext is actually in 73 million American homes, but only a few thousand people know it. Since 1983, CBS, NBC, and various local stations broadcast teletext pages. Lacking is a low-cost decoder to grab this material off the air and display it.

Videotext can provide an individual with an infinite number of pages, with the limit depending upon the capacity of the central computer's memory. In comparison to teletext, with a waiting time of up to half a minute to grab a frame, the response time with videotext is only a fraction of a second. On the other hand, videotext is much more expensive, involving telephone charges and, often, a small charge for each frame. Teletext is more one-way in nature than videotext, and somewhat less interactive. Videotext is a simplified version of computer time-sharing, allowing thousands of individuals to call up data from the memory of a central computer.

Teletext and videotext are both very limited in terms of the amount of information that can be displayed on a page or frame (usually only about a hundred words, or an equivalent amount of

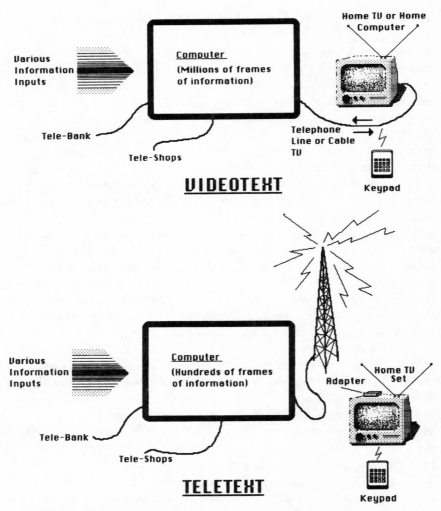

Figure 2–3. Videotext and Teletext Are Two Types of Interactive Information Services

Videotext is an interactive information service that allows individuals to request frames of information from a central computer via telephone or cable, for viewing on a video display screen (usually a home television receiver). *Teletext* is an interactive information service that allows individuals to request frames of information for viewing on a home television screen; the frames are transmitted in the vertical blanking interval of a conventional television broadcast signal. So in essence, videotext is wired teletext; the range of frames of information available to the user is almost unlimited with videotext. It is more highly interactive than teletext.

graphic material). A teletext frame in the U.S. holds up to twenty lines of text, with thirty-two characters per line. So teletext and videotext provide only small goblets of information, more in the tradition of radio than the newspaper (Smith, 1980, p. 249). The entire 400 frames available on a typical teletext system would correspond to only four pages of a newspaper.

The history of videotext and teletext is dominated by several European nations. The first videotext system, Prestel, was launched in England in 1979 by the British telephone authority (which is part of the British Postal Service). The original motivation for Prestel was to get a greater volume of message traffic on the underutilized British telephone system. At first, Prestel officials predicted they would achieve one million subscribers by 1981. They overestimated the actual rate of adoption; by 1984, there were but 20,000 home subscribers—this after the British government had invested $100 million in the system. The French and Canadians were also ahead of the United States in launching videotext systems, called Antiope and Telidon, respectively. Only in 1983 were the first videotext pilot projects launched by American firms (not by government agencies as in Europe).

Like videotext, teletext began in England. Engineers at the British Broadcasting System (BBC) and at the Independent Broadcasting Authority (the BBC's commercial counterpart) developed a television captioning service for the deaf in the early 1970s. This prototype soon evolved into a system similar to present-day teletext, which was launched in 1976: The BBC version is called CEEFAX, and the IBA's system is called ORACLE. Together, the two systems had 400,000 subscribers (about 3 percent of British households) in 1982 (Singleton, 1983, p. 128).

The French also launched their teletext system, calling it Antiope because it is compatible with the videotext system of the same name. Similarly, Canada's Telidon is a compatible system. Both the British and the French/Canadian teletext systems were tested in pilot projects in the United States.

Government involvement in videotext/teletext development can be a very expensive proposition. In 1978, the Canadian Department of Communications learned that Canadian cable and telephone companies were considering buying or copying the British or French systems. At that time, a marketing study in Canada had estimated that consumer spending for a computer-related technology would reach $8 billion by 1985. Accordingly, the Canadian government rushed the Telidon system, which had been designed by government R & D workers, into a pilot project in Ontario. In 1980, the Canadian government allocated $9 million, expanded to $40 million by 1983, to make Telidon available throughout Canada. Pilot projects were begun in four Canadian provinces, and, with an eye toward international sales, in the United States.

The Canadian government's strategy has met with some success. AT&T accepted the Telidon technology as one of the North American standards, along with Britain's Prestel and France's Antiope. The Times-Mirror's Gateway pilot project in the Los Angeles area uses Telidon equipment. On the other hand, Canadian consumers have been no more eager to adopt videotext/teletext services than have their U.S. counterparts; a 1982 projection of 40,000 Telidon sets in use fell far short, at only 400.

The videotext pilot projects underway in the United States are very expensive and will not return a profit for many years. Hence only the very largest media conglomerates are involved. Most of these corporations entering the videotext industry are in closely related media fields. Examples are the Knight-Ridder and Times-Mirror newspaper chains, and Dow Jones (a financial service that publishes the *Wall Street Journal*). IBM, Sears, and CBS are planning a joint videotext project called Trintex and RCA and Time are also considering one.

Despite this interest in videotext on the part of commercial providers, consumers have generally been reluctant to adopt. For instance, Knight-Ridder opened its Viewtron system in 1983 in Miami; it provides news reports, electronic mail, and home-banking and shopping. By early 1985, the Viewtron project had cost Knight-Ridder $35 million and only about 3,100 subscribers had signed up. Times-Mirror's Gateway project in the Los Angeles area cost $25 million, and by 1985 only had 500 subscribers. A household must buy a special receiver for $600 to $900 (or rent it for $20 to $40 per month) in order to subscribe to Viewtron or Gateway. Certainly this cost of adoption is a major reason for the discouraging subscription rates for these new videotext services. Several videotext services eliminate most of the initial expense by using a microcomputer for home reception, and this seems the logical way for videotext to go. Examples are the Dow Jones News/Retrieval, which began in 1974 and by 1985 had 200,000 subscribers to its financial data service; CompuServe, owned by H&R Block, the income tax service, had 190,000 subscribers; and The Source, owned by *Reader's Digest*, had 63,000 adopters in 1985. All three systems can be tapped by a home computer.

Knight-Ridder's Viewtron videotext system in Miami offers 7 million frames of news, stock market reports, weather, airline guides, the entire text of an encyclopedia, video games, children's math drills, sports scores, product evaluations from *Consumer Reports*, and even the police blotter at your local precinct. One can buy and sell stocks through Viewtron, purchase a variety of consumer goods, and send messages to anyone else who subscribes to the service. All this for an initial equipment purchase of $600 plus a monthly fee of $12 plus about $1 per hour for telephone company charges. For an information-freak, this is a real bargain.

Videotext promoters assumed that because people like to read newspapers, books, and magazines, they would transfer this reading habit to a new format, the videodisplay (CRT) terminal screen. Videotext has the special advantage of just delivering information you want without your having to leaf through all the rest of a newspaper, say, to get at the specific information in which you are interested.

Actually, there are a number of reasons why most individuals are "turned off" by videotext. For one thing, a videotext user can only read the television screen while sitting in one spot. In contrast, a newspaper can be read in a commuter train, while lying on the living room floor, etc. Also, it's difficult for videotext users to obtain the "big picture" of the day's news, so they don't know what they are chosing from (unlike looking at the entire front page of a newspaper). Someone has said that videotext is like looking at a football field through a drinking straw. Further, a problem for videotext (and teletext) is the prevailing attitude of many people that television is watched for entertainment, not for obtaining instrumental information.

Another consumer problem for the early videotext pilot projects, paradoxically, is that their range of content is so broad. Viewtron, for example, contains 300 pages of information, ranging from injury reports on professional football players, to the latest jokes. A more specialized content, with greater depth in each area, seems to be what consumers want. This lesson was learned by Britain's Prestel service, which in 1984 decided to concentrate on four areas: airline reservations, local news, home banking, and delivering computer software programs. Immediately after this change to greater specialization, the number of households subscribing to Prestel shot up, and the service began to turn a profit. Why did Prestel wait five years before specializing? Because it took that long to understand what potential consumers wanted and did not want. The general trend for most videotext systems in use today is to shift from delivering all types of information, to providing teleshopping and telebanking services (Noll, 1985).

What is the future potential of videotext in America? Industry experts estimate that by 1995 about 20 percent of all households in the U.S. will have adopted the service, about the same proportion that presently have adopted video cassette recorders. So videotext appears to be a new interactive medium whose rate of adoption is about ready to take off.

Teleconferencing: Electronic Meetings

Because of the capacity of interactive telecommunication to jump physical distance, one application of the new communication technologies is *teleconferencing,* defined as a small group meeting held by interactive electronic communication among three or more people in two or more separate locations (Svenning and Ruchinskas, 1984; Johansen and others, 1979, p. 1). This type of meeting saves the cost and burden of travel, which can amount to millions of dollars per year for a large corporation.

There are three main types of teleconferencing:

1. *Video teleconferencing:* a telemeeting in which video pictures are transmitted among the participants, who may be geographically far apart. Each set of participants usually come to a specially equipped conference room where they are seated around a table with microphones, television cameras, and a TV screen. During the teleconference they can show one another visuals—graphics, charts, and objects—and they can see whether other individuals are smiling, bored, or disapproving. Thus, video teleconferencing is much like a face-to-face meeting, although some of the visual context is eliminated (usually only a single camera is used). This is the most expensive and the least utilized of the various types of teleconferencing; it is also the most glamorous.

2. *Audio teleconferencing:* a telemeeting in which an audio channel (usually a telephone) connects the participants. Audio meetings are similar to a conference telephone call in which a special conference room equipped with high-quality voice transmission

technology is used. In a somewhat more costly version of audio teleconferencing, called audiographic teleconferencing, visuals may be transmitted among the distant participants in the telemeeting. Compared to face-to-face meetings, no nonverbal communication is possible in audio teleconferencing. The visual band has been eliminated from human communication.

Audio teleconferencing tends to be de-statusing because much of the context of communication is not visible: the $200 suit worn by an individual; the large, well-appointed office and its corner location on the top floor of the company's headquarters building.

3. *Computer teleconferencing:* a telemeeting in which a computer serves as a "meeting place" for a print-based exchange of messages among the distance participants (Johansen and others, 1979, p. 1). Computer teleconferences are least like face-to-face meetings because neither nonverbal, visual communication, nor "social presence," can occur. On the other hand, computer teleconferencing is asynchronous (and so the participants can "drop in" whenever they find it convenient to do so), able to provide a written record of the discussion, and capable of polling the participants on selected issues and tabulating the results (Johansen and others, 1979, p. 1).

The Hewlett-Packard Company has an electronic messaging system for transmitting information among its various plants and sales offices. The average cost is only about one cent per hundred-word message for domestic transmission, which compares very favorably with the cost of a first-class stamp in the U.S. (Pool, 1983a, pp. 190–191). The Hewlett-Packard system carries 25 million messages per year among its 45,000 employees.

One general issue explored in communication research on teleconferencing is the relative cost-effectiveness of the three modes compared to face-to-face meetings. The results indicate that computer, audio, and video teleconferencing (in that order) are more cost-effective than the average face-to-face meeting, mainly due to the considerable cost of travel that is saved, and the attendent savings in travel time. Video teleconferencing can not be justified just on the basis of time saved, but is a better alternative than face-to-face meetings on the basis of transportation costs (Short and others, 1976, p. 165). The participants in teleconferences perceive them as having greater formality; one advantage is a shorter meeting, but this advantage comes at the cost of losing much of the socio-emotional content that would be involved in a face-to-face meeting.

For instance, it is usually impossible to offer a cup of coffee or tea to a fellow teleconference participant (Short and others, 1976, p. 140).

Perhaps someday we may learn how to fulfill our socio-emotional needs in meetings via telecommunication. Or to get along without such socio-emotional aspects, should they turn out to be unnecessary for effective meetings.

SOCIAL PRESENCE AND NONVERBAL COMMUNICATION

Social presence is the degree to which a communication medium is perceived to be socio-emotionally similar to a face-to-face conversation. This quality is called social presense because it indicates the degree to which an individual feels that a communication partner is actually present during their exchange of information. Although they do not exactly define this concept, John Short and others (1976, pp. 66–76) review various research studies to show that face-to-face interaction is consistently regarded by respondents as highest in social presence, while a business letter or an audio-only channel is rated lowest, with visual media (such as a video teleconference) perceived as midway in social presence between the extremes of audio media and face-to-face communication. Thus, the degree of social presence varies directly with the degree of band width (a video signal requires a frequency band 600 times wider than a radio signal, the equivalent of a tire track compared to a pencil line). The communication media high in social presence are those in which nonverbal as well as verbal communication occurs. Is there anything to social presence other than just nonverbal communication? We wonder exactly what it might be.[4]

Nonverbal communication can occur only when two or more individuals are actually in each other's presence, or when they are linked by a visual medium (such as video). Nonverbal communication serves to complement verbal communication in some very important ways:

- Such nonverbal acts as head nods and eye contact provide evidence of a listener's continuous attention during a conversation.
- Who shall speak and for how long, as well as other aspects of channel control, are mainly determined by the nonverbal band of communication. Without such nonverbal clues, as in a long-

distance telephone conversation by satellite, both individuals often begin talking at once.

- Continuous feedback to the speaker is usually provided via nonverbal channels; otherwise a speaker must wait for a spoken reply to indicate the listener's feedback to his previous remark.
- The socio-emotional aspects of building and/or maintaining an interpersonal relationship are mainly conveyed by nonverbal communication. Could you imagine trying to get to know someone that you never saw, or never met in person?

Our discussion of the socio-emotional content of computer communication might have implied that little or no emotional content is present. This is not so. For instance, a participant in one computer bulletin board said: "It's nice to have someone to talk to when you want to be alone" (Phillips, 1982). Later (in Chapter 6), we will present evidence from research studies to indicate that computer communication often contains considerable socio-emotional content, sometimes as much as 30 percent of all the messages that are exchanged.

What do we mean by the concepts of socio-emotional and task-related communication? These terms come from a Harvard University social psychologist, Robert Bales (1950), who categorized a dozen group member roles such as information-provider, opinion-requestor, and tension-reliever. He began this approach in a study of the group meetings of a local chapter of Alcoholics Anonymous, where he found it difficult to ask direct questions of respondents in a survey. In the heyday of small group communication research, observers were trained to identify these roles when they were played in a group; then the data on these categories could be analyzed to better understand the nature of group communication. *Task-related communication* asks for information, gives information, asks for suggestions, gives suggestions, asks for opinions, and gives opinions. These roles reflect communication content that generally can help directly move a group, an organization, or a system toward its objectives. *Socio-emotional communication* shows solidarity, tension relief, agreement, antagonism, tension, and disagreement. These roles reflect communication content that generally can facilitate or inhibit the process through which a group, an organization, or a system passes in carrying out its activities.

The two analytical concepts almost fell into complete disuse in the late 1970s. But they became very useful again in the 1980s as tools for the investigation of all types of computer communication.

The new media make it easily possible to communicate with someone without being physically present. This fact raises the important question of the importance of nonverbal communication. When does nonverbal communication have to be present and when can effective communication occur without it? Only partial answers are available now from communication research. For example, "getting to know a stranger" is considered to be an inappropriate situation for using interpersonal telecommunication (Short and others, 1976, p. 139). So is bargaining and negotiation.

An important future research question to be answered is how human communication is changed when it occurs via computer. Certainly one distinctive quality is the deemotionalization in keyboard-to-keyboard communication. Human emotions are difficult to convey in telegraphic writing. Imagine trying to tell a joke to another person by computer; some people type the words *ha, ha* at the appropriate place. That hardly seems to convey the full meaning of laughter. The reason why socio-emotional content is difficult to convey by computer is the limited communication band that is being utilized (only a print channel). Nonverbal communication cannot occur when a visual sense of one's communication co-participant is missing. Of course, audio communication is also missing when messages are exchanged by computer; this means that rapid feedback cannot occur. In short, computer communication is about as far from face-to-face interaction as one could imagine, and thus is very low in social presence.

The New Cable TV

By the mid-1980s, 32 million (40 percent) U.S. television households subscribed to cable television, and this rate was climbing steadily. Most European nations had an approximately similar rate of adoption, with the Netherlands and Belgium at 60 to 70 percent adoption, as was Canada (Pool, 1983a, p. 155). About 60 percent of American households were passed by cable in the mid-1980s, and most of these will eventually subscribe. In fact, about 70 percent of American households are expected to adopt cable TV by the year 2000.

Where did cable TV come from? And what forces drove it to become so popular in recent years?

Cable television started in 1950 at Lansford, Pennsylvania, and flying over this area, one can easily understand why. A series of small mountain ranges run from north to south in Pennsylvania with small cities like Bethlehem, Wilkes-Barre, and State College located in the valleys. Most television signals are broadcast either from Philadelphia in eastern Pennsylvania or from Pittsburgh in the western end of the state. Television waves do not bend, and so people living in the Pennsylvania valleys could not receive broadcast signals. So a local businessman in Lansford put an antenna atop one of the nearby mountains and ran a cable into the homes in the valley—thus was cable TV born.

During the 1950s and 1960s, cable television was perceived as a means of obtaining improved TV reception in remote rural areas. Only 12 percent of American households were hooked up by the early 1970s. They usually received the three television networks and perhaps one or two independent educational stations. Cable TV didn't amount to much.

Then, in 1975 entrepreneurs got the idea of delivering television programming to local cable systems by satellite. One was HBO (Home Box Office) and the other was Ted Turner, a professional-sports mogul and owner of an Atlanta TV station. Now, an almost unlimited variety of programming can be obtained when you hook up your home to a cable TV system: In addition to ABC, CBS, and NBC network television, there is a twenty-four-hour news channel, a twenty-four-hour sports broadcast, a religious channel, recent Hollywood films, a health channel, etc. By 1985, there were forty satellite-delivered television programming services. With this channel diversity becoming available, the rate of adoption of cable TV suddenly took off: up to 40 percent of American households by 1985 (Figure 2–4).

Given satellite delivery of programming, what really began to drive cable television was pay-TV (Figure 2–5). For an extra fee each month (usually around $10) a household can receive a channel showing recent movies, the Playboy Channel (soft porn), etc. (An added attraction of pay-TV is the absence of commercials.) The desire for access to these pay television services motivated an increasing number of American households to hook up to cable. In fact, as the rate of adoption of pay cable went up, up, up, the adoption of

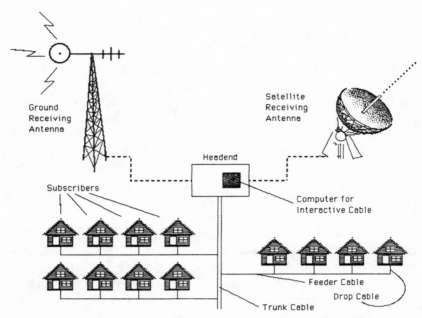

Figure 2–4. A Cable Television System, Showing an Antenna for Receiving Satellite Programs, and a Computer for Interactive Services

In 1975, 12 percent of American households had adopted cable TV, which was then provided by 3,500 cable TV systems. HBO's strategy of using satellite transmission to deliver television programs to cable systems inaugurated the widespread availability of pay-TV. The extra income generated by pay-TV motivated cable operators to build new cable systems in large cities, where construction costs had been prohibitive. By 1985, 40 percent of American households had adopted cable television, which was provided by about 5,000 cable systems. So the new technology of satellite-transmitted programming created greater economic profitability for cable operators and an attractive new cable TV content (first-run movies on pay-TV channels) for consumers.

Interactive cable television allows for sending text and graphic frames, as well as full video pictures, to home television sets via cable, in answer to request. In 1985, less than 1 percent of American households had interactive cable.

cable TV went up with it. The extra income generated by pay-TV for cable operators enabled them to build cable systems in large cities, where construction costs had been prohibitive. Particularly popular with American households who sign up for pay-TV is the opportunity to view recent Hollywood films and this provided the big push to the recent growth of cable TV in the United States. Pay-TV became an alternative to movie attendance in theaters. The im-

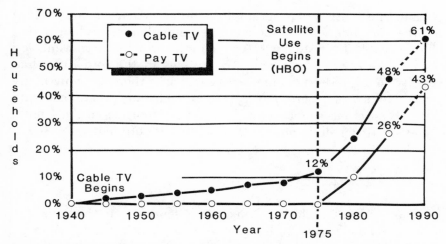

Figure 2–5. The Rate of Adoption of Pay-TV Gave a Boost to the Rate of Adoption of Cable TV

The data on rate of adoption shown here suggest that the post-1975 boost in cable television adoption is due in large part to the effect of the growth of pay-TV among American households. With the greater diversity of channels available on cable television (including the pay channels) due to satellite transmission, the rate of adoption of cable television surged.

portance of feature films in the rapid growth of cable TV is also demonstrated by the British experience, as well as that in several other countries.

HBO is owned by Time, Inc., a news magazine company that was convinced the future of cable TV was in providing feature films, especially Hollywood blockbusters. HBO started in 1972, and grew very slowly for a couple of years because it was unable to rent really top-rated films, and because it delivered its programs to local cable systems by video cassettes or by microwave, which was slow and unwieldy. Time, Inc., was losing a lot of money. Then the corporation decided to marry cable TV with another technology, satellites, as a means of delivering HBO programs to the cable TV systems that purchased the service. Satellite delivery was an expensive venture; the board of directors of Time, Inc., committed $7.5 million for leasing a satellite transponder from RCA for five years. The satellite transmissions began in 1975, and with it pay cable began to take off, especially when HBO began to provide current box-office hits. The high profits racked up by HBO soon attracted a host of other pay-TV services, also delivered to local cable TV systems by satellite.

Satellite delivery of programs to cable TV systems means that the number of channels is much greater than before. While a ten-channel cable TV system was typical in the presatellite transmission days, some of the new cable TV systems in the mid-1980s had seventy to a hundred channels. Even though there are now many more channels of programming, is there really any greater program diversity? Actually, not so much. Ethnic minorities in the United States say they still do not see themselves very often on cable TV.

SATELLITE COMMUNICATION

In 1945, Arthur C. Clarke, then a flight lieutenant in the Royal Air Force and now known for his science fiction writing (*2001: A Space Odyssey*), proposed the idea of a communication satellite in the British academic journal, *Wireless World*. Clarke argued that a geosynchronous satellite, positioned 22,300 miles above the equator so that it would be stationary to a given point on the Earth's surface (even though moving through space at 7,000 miles per hour), would be the perfect platform for television broadcasting. The television signal would cover about one-third of the Earth's surface (its "footprint"), and could be particularly advantageous in mountainous areas and in nations with a large land area, such as Russia, China, Brazil, and the United States. In essence, satellite communication could remove the effects of distance on transmission costs.

For several years, Clarke's suggestion did not lead to action. Then, in 1954, John R. Pierce, a noted engineer at Bell Labs and a science fiction writer himself, became the leading force in realizing Clarke's fantasy. Launch of the Russian sputnik in 1957 set off a "space race" between the United States and the Soviet Union, leading to more and more powerful satellite launching vehicles (basically these were missiles). The first U.S. communication satellite, WESTAR 1, was launched in 1974. By the late 1970s, communication satellites were being utilized not only to transmit television broadcasts, but also for long-distance telephone calls (perhaps you have noticed the slight delay of a few seconds caused by the 44,600 miles traveled by an Earth-satellite-Earth transmission), as well as a variety of other communication tasks. The communication satellites were positioned 4° apart (about 1,000 miles) over the equator so as to prevent interference, as only about 75° of arc lies over the North American continent (due to more advanced technology, satellites can now be positioned only 2° apart). Only a limited number of satellite "parking spaces" are available for U.S. satellite broadcasting. A satellite has from twelve to twenty-four "transponders" (transmitters-responders), each of which is a device that receives a signal from Earth, converts it to another frequency, amplifies it, and beams it back to Earth (Figure 2–6). One transponder can handle 1,300 simultaneous telephone calls, twelve radio station signals, or one color television signal (Singleton, 1983, p. 83).

Use of a transponder can be leased for a year for about $1 million; a circuit on a transponder can be leased for $15,000 or so. Building and launching a communication satellite costs about $75 million. Satellite communication has become a huge commercial enterprise. Clarke wrote a history of communication satellites, which he ruefully entitled: *A Short Pre-History of Comsats; or, How I Lost a Billion Dollars in My Spare Time.* He did not patent his idea (perhaps it could not have been patented); in any event, it has proven to be a very powerful one. Along with computers and new transmission technologies such as fiber optic cable, satellites are fundamental to the new communication systems of the 1980s. Because

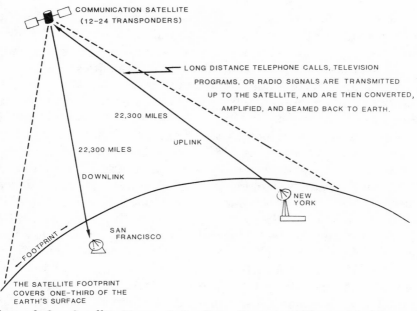

Figure 2–6. Satellite Transmission Is an Important Element in Many of the New Communication Systems

satellites are relatively invisible to the general public, their role in the Communication Revolution has not been as fully appreciated as that of computers, but a local cable TV system typically pays 10 to 20 cents per subscriber per month for each satellite programming service, and this charge is passed along in the subscriber's monthly fee.

Among the major, direct impacts of communication satellites are:

- Much cheaper long-distance telephone service
- A take-off in the rate of adoption of cable TV in the United States, as satellite transmission of television programming enriched the channel selection that was available, especially pay-TV programming
- Facilitated national newspapers in the U.S.—such as the *Wall Street Journal, The New York Times,* and *USA Today*—who transmit their pages by satellite to scattered printing plants around the country
- Speeded-up the flow of international news, so that film of a news event happening on the far side of the world can be viewed, same day, on television network news.

Because satellite communication eliminates the cost of greater distance, this technology has the potential to decentralize all types of information-intensive institutions—for example, banking, research, and government. Instead of crowding these facilities into a central city, satellites make it possible to scatter these types of employment closer to workers' homes. During the 1980s, New York, San Antonio, San Francisco, and other U.S. cities launched teleports, "satellite antenna farms," in which these information facilities are concentrated. Urban officials hoped that these information harbors will have an effect on city growth similar to that of seaports in the past.

Communication satellites also make possible direct broadcast satellite (DBS) systems, enabling individuals to receive satellite-delivered television directly in their homes through a small dish-shaped antenna that feeds their television receiver. DBS is television without a middleman, skipping local TV stations and cable systems. The cost of adopting a rooftop antenna system in the mid-1980s was only $2,500, and dropping.[5] The number of dish owners shot up from a few thousand in 1979 to about a million in 1985. This rapid rate of adoption was also due to a 1984 federal law making it legal for a private citizen to own and use an antenna to receive satellite television transmissions; in response, TV programmers are scrambling their signals. Many of the adopters of the giant upside-down mushrooms live in remote rural areas, which are infeasible for cable TV to reach, but an increasing number live in suburban areas and want greater diversity in their television program reception. With one's dish antenna pointed up at each of thirteen broadcasting satellites, an individual can receive 120 channels including everything from free HBO and every televised football game in the nation to the "live feeds" of network TV programs (enabling dish owners to watch national newscasters kibbitz during commercial breaks).

In mid-1985, a wide variety of TV programming could be picked out of the air by someone with a backyard satellite dish: On SAT-COM F4, the National Christian Network is just five transponders away from the Playboy Channel, and in between are Netcom (a teleconferencing network), Sportsvision, American Movie Classics (films that are at least fifteen years old), and Home Sports Entertainment. Forty degrees west is WESTAR 5, containing an X-rated network, the Financial News Network, a Pennsylvania harness racing service, a Michigan sports network, and an evangelical religious network. Eleven degrees further west is GALAXY 1, carrying HBO, Cable News Network, Spanish International Network (featuring Mexican soap operas), C-Span (the network covering the U.S. Congress), and the Entertainment and Sports Programming Network. So it goes.

In addition to satellites, a major improvement in communication transmission during the 1980s is provided by fiber optic cable. Compared to the copper wire utilized in most present communication systems, fiber optics are much cheaper and have important technical advantages. Fiber optic cable transmits messages through a tiny glass tube as light pulses (that is, light beams) that are generated either by lasers or light-emitting diodes. Fiber optics are beginning to replace copper wire as a means of transmitting telephone and other communication messages. The main advantages are:

1. Efficiency: As many as 40,000 telephone calls can be carried through a bundle of optical fibers the size of your finger. Although it is much smaller than copper wire, fiber optic cable is flexible and can be spliced.

2. Two-way: Fiber optic cable can simultaneously transmit messages in both directions without them interfering with each other. Naturally, this quality is very important in two-way interactive cable TV systems.

3. Noninterference: Because fiber optic cable carries *light* messages, rather than *electronic* impulses, fiber optic transmission does not interfere with radio, television, or other types of electronic communication. So, fiber optics free communication systems from cross-talk, leakage, and other noise problems found in electronic communication.

The "wired cities" pilot projects in Japan, France, and other nations use fiber optic cable to interconnect homes and businesses in a local community (Dutton and others, in press). In order to handle the much higher rate of telephone traffic during the 1984 Olympics in Los Angeles, fiber optic cable was installed. It is also being utilized to replace copper telephone wire in such places as the eastern seaboard section of the United States, where a high volume of telephone communication occurs.

Qube in Columbus*

On December 1, 1977, an interactive cable television system, Qube, began operating in Columbus, Ohio. Promotional material for Qube stated: "The age of passive viewing is over," and "Touch the button. Qube brings the world to your fingertips" (Becker, 1984). Qube's inauguration was the first time a U.S. audience was provided a two-way link with commercial television programming.[6] Qube became feasible because computer costs were falling sharply due to technological advances in the increasing miniaturization of semiconductor chips. Thanks to its computer component at the head-end of the cable system, Qube was interactive in two ways: first, Columbus householders on Qube sent back signals to the system, when prompted for this feedback, through a hand-held keypad with keys numbered from "0" to "9," and second, the computer regularly polled each Qube household as to what programs were being watched. This more passive type of interactivity was very valuable to the Qube system in providing a sophisticated data base about each household's program preferences, but Qube management publicly stressed the user-initiated interactivity and seldom discussed the computer-monitored type of interactivity. One can imagine the privacy issues surrounding head-end computer records showing you viewed an X-rated program. Qube subscribers had signed an agreement for such computer-recorded data to be gathered, with the understanding that it would not be released on an individual basis to parties other than the owners of the Qube system, Warner-Amex. Actually, all of these data were hoarded carefully by Warner-Amex; as a result, scholars and the public know little about the use behavior of Qube's subscribers.

In 1984, the Qube system cut back substantially on the number of interactive programs being provided, presumably an indication that the interactive programs were not profitable. Why didn't Qube live up to its original expectations? One reason was the very rapid diffusion of home computers, which, with a modem and telephone, can provide access to electronic messaging systems, computer bulletin boards, large data bases such as The Source and CompuServe, etc. Thus, microcomputers provided a type of competition for interactive cable TV that was not anticipated when Qube was launched.

How popular was Qube with the citizens of Columbus? In 1984, after seven years of operation, the Warner cable TV system in Columbus passed about 105,000 homes; 60 percent subscribed to a ten-channel basic cable TV service, and 75 percent of these also subscribed to Qube (for a small extra monthly charge). This is not an overwhelming vote in support of interactive cable television, but these data certainly suggest that it has a great number of fans.

On the other hand, there is evidence that interactivity has never been very important to Qube's subscribers (Becker, 1984): (1) only a couple of percent of the Qube households watch even the most popular interactive programs (a game show in which home viewers actively participate and a political policy program), and (2) only 33 percent of Qube subscribers each week participated at least once in any kind of interactive program. These low use rates of interactive programs are all the more discouraging given the ingenuity with which Qube's rather limited interactive capability was utilized in certain programs. For example, on "Home Book Club" a panel of experts discussed a new book each week that had been chosen by Qube participants the previous week; the topics discussed were also determined by viewers of the "Home Book Club" program. In another example, the citizens of Columbus participated in discussions of a planning report prepared by the city's professional planners; Qube householders could indicate when they wanted to ask a question of the planning commission members, then they were called by telephone at home, and their question was answered on the air.

Warner-Amex, owners and operators of Qube, is a joint venture of Warner Communications (the media conglomerate) and American Express (the giant credit card and banking corporation). Amex's motivation for its joint ownership of Qube was in order to learn more about the public's acceptance and use of telebanking and teleshopping services. What did Warner-Amex get out of the Qube experiment in Columbus? A loss of over $30 million over the first seven years—the experiment was highly unprofitable in a direct sense. However, the operation of Qube helped Warner Cable establish its reputation as a technologically sophisticated cable company. This image helped Warner win the "cable wars" in Cincinnati, Dallas, Houston, Milwaukee, Pittsburgh, and St. Louis; once these cable franchises were won, Warner Cable began to deliver ninety minutes of interactive cable television programs every evening to a total audience of about 350,000 households. (In 1984, this programming was cut down to a bare minimum because Warner could not find advertisers to support it. Evidently these interactive programs were not attracting large enough audiences to interest advertisers.)

Perhaps Qube in Columbus should best be considered a huge and expensive experiment, that provided its sponsors with a sophisticated database about household use of interactive cable television.

WIRED CITIES[7]

By no means do Columbus, Ohio, and the other Qube cities in the U.S. have the most advanced interactive cable systems in the world. Three other "wired cities" have systems that are technologically much more advanced. Higoshi-Ikoma, Japan; Biarritz, France; and Milton Keynes, England, provide a somewhat utopian view of what the interactive communication technologies can do to promote a sense of community. Although none of these systems are profitable (in fact, each project represents a huge investment with only a small financial return), these three wired cities are important laboratories for better understanding of the adoption and impacts of interactive communication technologies.

All three of these cities are relatively elite, with a high proportion of well-educated, higher-income individuals who are concentrated in technological occupations (for example, scientists and engineers). Higoshi-Ikoma is a suburb between Osaka and Kobe, in which 156 households in 1978 were connected by fiber optic cable to a head-end computer. Each home is equipped with a television set, a keypad for expressing opinions by voting on a limited set of choices, and a camera that is fixed to the top of the TV set making the Hi-Ovis (highly interactive optical visual service) system two-way audio *and visual*. A red light alerts the individuals in a household that their camera is on; they can place a lens cover over the camera anytime they wish, so one's *1984* fears about the Higoshi-Ikoma project are in part allayed. The project was funded by the Japanese Ministry of International Trade and Industry (MITI) in order to demonstrate the usefulness of the interactive cable technology so that it could be sold to other nations. Televoting on issues (with the keypad) is popular in Higoshi-Ikoma, but not use of the home video camera to send messages (respondents say they want to clean their house before it is shown on TV, men say they feel they should shave first, they can't drink beer while watching TV out of fear of being seen on camera by their neighbors, and housewives say they feel too nervous to talk on TV). Evaluations of Hi-Ovis use patterns indicate that the interactive, two-way capability of the system is utilized a great deal by a few enthusiasts versus very little by a majority of relatively indifferent individuals. A generally similar pattern of use seems to characterize interactive communication systems in other nations, particularly when the new capability requires a certain degree of complex or unusual behavior that has to be learned in order to utilize the system.

Milton Keynes is a new town on the distant outskirts of London. In 1982, its 20,000 households were offered a coaxial fiber system for interactive television. Sponsor of the Milton Keynes project is the British Postal Service.

Biarritz is a small resort town with a population of 30,000 located on the Atlantic coast in southwestern France. A fiber optic cable system was adopted in 1984 by 1,200 households and 300 businesses and government agencies. The Biarritz project is sponsored by the French Ministry of Posts/Telephone/Telegraph. Each user of the system receives a piece of equipment that combines a telephone, microcomputer, and color TV set. It can be used as a picture telephone, for cable TV, to access the national videotext service, and for teleshopping. Why is the French government willing to invest several million dollars in the Biarritz project? Because French officials realize that their nation, in order to succeed economically in an Information Society, must lead the world in new communication technologies. The Biarritz technologies are expected to create 30,000 new jobs over the next fifteen years through the sale of the technological equipment; this new employment is needed to replace jobs that will be lost in the telephone manufacturing industry (as almost all French homes now have phones). So in this sense, the Biarritz project is a kind of sales demonstration for French high-technology.

Notice the advantages and disadvantages of the role of government in the Qube project versus the three wired city projects. Qube is conducted as a joint venture by two private corporations; the lessons learned about acceptance and social impacts are the sole possession of Qube management, and need not be shared with the scholarly community or with Qube's possible competitors. The $30 million loss is also the sole responsibility of Warner-Amex.

The Japanese, English, and French wired cities projects would not have happened without government sponsorship. Private firms are involved in designing and producing the interactive technology equipment, and, in turn, may profit from future sales if the pilot project in their nation is successful. The national government may gain through the future generation of jobs and the income taxes that will thus be paid. Because these three experiments are public, they are usually evaluated by communication researchers from universities or market research firms, and the results are widely distributed.

Summary

The six communication technologies just discussed in this chapter are compared as to interactivity, de-massification, and asynchronicity in Table 2–2.

Table 2–2. A Comparison of the New Communication Technologies on the Basis of Interactivity, De-Massification, and Asynchronicity

New Communication Technologies	Degree of Interactivity	Degree of De-Massification	Degree of Asynchronicity
1. Computer communication through computer bulletin boards and electronic messaging systems	High	High	High
2. Videotext	High	High	High (?)
3. Teletext	Medium	High	High (?)
4. Teleconferencing			
(a) Video	High	High	Low
(b) Audio	High	High	Low
(c) Computer	High	High	High
5. Basic cable TV (with satellite programming)	Low	Low	Low
6. Interactive cable TV	High	High	High

Notes

1. Sheizaf Rafaeli (1985, pp. 53–54) found a rather surprising degree of feedback from the public to mass media institutions. One-third of the respondents in a San Francisco area survey said they had never written to the editor of a newspaper, and 20 percent said they had never called

in to a radio or television station in order to be on the air (as in a radio talk show).

2. The following several sections are based directly on Everett M. Rogers and Sheizaf Rafaeli (1985).

3. The terminological confusion about what to call videotext and teletext traces from a lack of standardization of terminology, the fact that rival systems were launched in different nations, and because of the changing information technologies underlying videotext and teletext. Some individuals insist there should be no final "t" in videotext. Others prefer to use "viewdata" instead of videotext. Some use videotext for all two-way systems, and teletext for all one-way systems. Still others use videotext as a generic term for all interactive information systems, both videotext and teletext. Are you confused?

4. Further, we wonder how distinct the concepts of social presence and interactivity are, as each is defined as being similar to face-to-face communication.

5. In 1975, a satellite Earth station cost a cable TV operator about $125,000, in part because the Federal Communications Commission mandated that an enormous dish about thirty feet in diameter was necessary for high-quality reception. By 1985, antenna technology had been improved so that a smaller dish fifteen feet in diameter (and costing only $5,000) provided equally high-quality reception to the much larger dishes of a decade previously. The price continues to drop.

6. Previously, however, three evaluations of interactive television for educational purposes had been conducted in the mid-1970s. These studies were sponsored by the National Science Foundation and carried out in Reading, Pennsylvania; Spartansburg, South Carolina; and Rockford, Illinois. The general picture that emerged from these three pilot projects was that interactive cable television had little relative advantage over noninteractive media in educational applications (Baldwin and others, 1978; Brownstein, 1978; Burns and Elton, 1978; Clarke and others, 1978; Elton and others, 1978; Fredin, 1983; Kay, 1978; Kwiatek, 1982; Lucas, 1978; and Moss, 1978).

7. Much of the material in this section is drawn from the author's visits to each of the three wired cities in recent years, and also from a conference on the wired cities held by the Washington Program of the Annenberg Schools of Communications in Washington, D.C. A book entitled *Shaping the Future of Communications: National Visions and Wired City Ventures* is being edited by William Dutton, Kenneth Kramer, and Jay Blumler as a proceedings of this conference.

CHAPTER 3

History of
Communication Science

"Communication has developed like a scholarly discipline. But has it produced a central, integrated body of theory on which the practitioners of a discipline can build and unify their thinking? I am afraid it has not."

Wilbur Schramm, 1983

In order to understand the full nature of the new communication technologies, one must look to the past as well as into the future. An important part of the relevant history of communication technology is the rise of the academic field of communication science. This discipline grew to intellectual strength mainly in the United States during the period following World War II, although its roots trace back another thirty years or so to European and American foundations.

In this chapter, we trace the rise of communication science as an academic discipline over the past several decades, taking up each of the main intellectual contributors. Our theme is to show how each of the communication technologies that came on the American scene— press, film, radio, and television—were related to the important events, ideas, and individuals in communication research. We shall carry our historical analysis up to the 1980s era of interactive communication (whose history is handled in other chapters).

A *Personal Perspective*

The present chapter tells the history of communication research from the personal viewpoint of the author; this subjective account helps make the intellectual history of communication research more interesting and logical to follow, although it necessarily injects my own viewpoint into the account. I began graduate work in 1954, shortly after the important "turning point" for communication science in 1950, and thus my career has coincided with much of the history that I describe here. Other than the present analysis, the history of communication science can be found in only rare and very partial accounts, such as chapters by Wilbur Schramm (1985) and Daniel Czitrom (1982). This lack of historical attention is strange, considering how long communication research has been around and how important it has become.

I received a doctorate in sociology in 1957, about the time the first Ph.D.s in communication were being awarded by Schramm at the University of Illinois. I had earned my master's at Iowa State University, and considered enrolling in communication at Illinois, but decided instead to remain in Ames. Those were exciting times in the new, wild field of communication research. Important books were appearing: Schramm's *The Process and Effects of Mass Communication* (1954), Norbert Wiener's *Cybernetics* (1948) and *The Human Use of Human Beings* (1950), and Claude E. Shannon and Warren Weaver's *The Mathematical Theory of Communication* (1949). At the time, we did not realize the role that this latter book would play in laying out the basic paradigm for the field of communication science. Those of us who were graduate students in the social sciences in the 1950s recognized that communication was a very fundamental human process in explaining behavior change. This viewpoint was being expressed by a small set of scholars who were coming together from different disciplinary backgrounds around a common model of communication, the one proposed by Shannon and Weaver.

Works by the four founding fathers of communication science influenced me in the mid-1950s. Paul F. Lazarsfeld's book with Elihu Katz, *Personal Influence* (1955), was discussed in graduate seminars. We read Kurt Lewin's books on group dynamics and I remember traveling to Iowa City, where Lewin had taught, for discussions about his theories with those of his disciples still at the University of Iowa. I was aware of Harold Lasswell's dictum, "*Who says what* in *which channel* to *whom* with *what effects?*" We studied Carl Hovland's persuasion research being conducted at Yale, building upon his wartime experiments on the communication effects of military training films.

I assumed that quantitative research methods offered the most promising approach to understanding the nature of human behavior change, so I minored in statistics at Iowa State, the university where the field had been launched in the United States by the English agricultural statistician R. A. Fisher in the 1930s. My dissertation dealt with the diffusion of agricultural innovations among farmers in a rural Iowa community. Mainframe computers were then (in 1957) becoming available for use in scholarly investigation, and my dissertation included a multiple regression; perhaps it was a sign of those times that I checked the computer print-out with a desk calculator.

Thus my graduate education was empirically oriented and based on quantitative methods, all grounded in an understanding of the communication theory of that day. After completing my doctorate, I moved to Ohio State University as an assistant professor of rural sociology, specializing in the study of the diffusion of innovations. At Ohio State certain faculty members in speech, journalism, psychology, and sociology formed an interdisciplinary graduate program in communication research, and I participated actively in it.

After six years in Columbus, I moved to a faculty position at Michigan State University in the Department of Communication, which was, in the mid-1960s, enjoying its golden years. Its professors and students were seeking to define the field of communication science. The department was orchestrated by David Berlo, one of Schramm's first Ph.D. students at Illinois. Nine years later, I moved to the University of Michigan, then for a decade to Stanford (where I had been offered Schramm's chair after his retirement), and finally to the University of Southern California in 1985. During this time, my interest in interactive communication technologies grew stronger.

European Roots: Tarde and Simmel

American science got where it is today in part because of a tremendous intellectual transfusion of academic talent from Europe, begin-

ning in 1933 with the Nazi party takeover in Germany. Jewish professors were dismissed from their university positions, and several
thousand of the top scientific minds migrated to the United States:
Albert Einstein, Erik Erikson, John von Neumann, and many
others. These émigré scholars contributed to military research during World War II; for instance, many of them worked on the
Manhattan Project while Germany's atom bomb project floundered
(Coser, 1984).

The Europe-to-America intellectual migration of the 1930s directly benefited the founding of communication science in the United
States. Many key thinkers in communication research (for example,
Kurt Lewin and Paul F. Lazarsfeld) were émigrés, and most of the
American-born had studied for their Ph.D.s in European universities. So while communication science started in the United States,
it had strong roots in Europe.

Figure 3–1 is a map of the history of communication science,
showing the dozen or so individuals that played particularly key
roles in this development. Note that the origins of this intellectual
history occurred in Europe in the late 1800s. At that time, European
universities, especially those of Germany, were the best in the
world. European social scientists of this era indirectly influenced the
communication science that was to arise later in the U.S.: Max
Weber, the great German scholar of bureaucracy; August Comte,
the father of sociology, and Emile Durkheim, a pioneer in using empirical research methods, both of France; and Sir Herbert Spencer,
known for social Darwinism. In addition, there were two more-
direct influences on American-style communication science: Gabriel
Tarde of France, and Georg Simmel of Germany.

Tarde was a French judge, and he based many of his sociological
observations on the human behavior that came before him in his
courtroom. Tarde set forth a theory of imitation, that is, of how individuals were influenced by the behavior of others with whom they
came in daily contact. These understandings about imitation provided a basis, forty years later in the United States, for communication research on the diffusion of innovations (Rogers, 1983) and on
social learning theory (Bandura, 1977). *Diffusion* is the process by
which an innovation is communicated through certain channels
over time among the members of a social system. Tarde observed
that the rate of adoption of a new idea usually followed an S-shaped
curve over time: At first, only a few individuals adopt a new idea,
then the rate of adoption spurts as a large number of individuals ac-

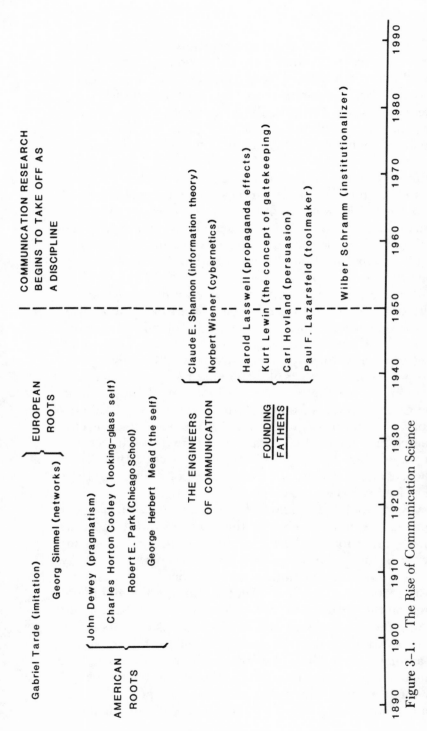

Figure 3-1. The Rise of Communication Science

These thirteen scholars played a crucial role in the rise of communication during the half century leading up to its take-off around 1950, when the Shannon model of communication became the widely accepted paradigm in the field. The four American "roots" emphasized the subjectivity of human communication, a viewpoint that was deemphasized when the four founders carried communication research into a strong empirical tradition. Wilbur Schramm's unique contribution was to achieve the institutionalization of communication schools in universities.

cept the innovation, and, finally, the adoption rate slackens as only a few individuals are left to adopt. Astutely, Tarde recognized that the S-curve "took off" when the opinion leaders in a system adopted the new idea.

Georg Simmel was a father of social psychology, the study of group influences on individual behavior. His book, *The Web of Group-Affiliations*, was written in 1922 but not translated into English until twenty-four years later (Simmel, 1946). This volume introduced the theory of *communication networks*, which consist of interconnected individuals who are linked by patterned flows of information (Rogers and Kincaid, 1981, p. 63). To Simmel, the essential question for understanding human behavior change was "to whom is the individual linked by communication ties?" Simmel provided the theoretical stimulus for studying communication networks, but empirical investigations on this topic were not to occur for several decades, until Jacob L. Moreno (1934) demonstrated the appropriate research methods for measuring and analyzing networks. He described communication networks as the "kitchens of public opinion."

Simmel and Tarde were theorists of communication whose ideas were not converted into empirically tested propositions until years later, on the other side of the Atlantic, where communication research methods were emphasized, and where the empirical approach to communication research flourished.

Four American Roots

Four turn-of-the-century American scholars with important influence on communication science were John Dewey, Charles Horton Cooley, Robert E. Park, and George Herbert Mead. Each contributed seminal ideas to the social science of human communication. Each of these four "roots" placed communication at the center of a conception of human behavior, although today none of the four is regarded mainly as having been a communication scholar: Dewey is considered a philosopher and psychologist; Cooley a sociologist of personality socialization; Park the key figure in the Chicago School of Sociology; and Mead a social psychologist. All four were highly philosophical, although they each utilized some kind of data as a basis for their theories. They were not experimenters or survey researchers, however.

Dewey, Cooley, Park, and Mead all stressed a phenomenological approach to human communication, emphasizing that the individual subjectivity of how a message is perceived is an essentially human quality. Thus, to these first four American scholars of communication, how an individual makes sense out of information, and thus how meaning is given to a message, was a fundamental aspect of the communication process. This subjectivism was recognized but deemphasized in the linear models of mass communication that came in the 1950s and thereafter, and the four roots of communication science have been generally ignored in contemporary times. That is a mistake, and it is serious.

Dewey, Cooley, and Park were called the Progressive Trio by Czitrom (1982, p. xii) in his history of the field of mass communication. This label derives from the great faith these individuals had in mass communication (especially newspapers, the dominant mass medium of their day) as an agent for restoring a moral and political consensus to American society. Industrialization, urbanization, and massive migration from Europe to the United States caused social problems of deep concern to the Progressive Trio. They hoped that the new mass communication technologies of their day would assist in the amelioration of society. Their blind spot was in overlooking the yellow journalism, newspaper barons (Hearst and Pulitzer), and commercial advertising that suggested the media were themselves very imperfect.

JOHN DEWEY: PRAGMATISM

Like the other three American roots of communication science, Dewey came from a liberal, small town and Protestant background. During his early years of teaching philosophy at the University of Michigan (from 1884 to 1894), Dewey influenced both Cooley and Park, infecting them with a view of mass communication as a tool of social change. Dewey hoped to "transform philosophy somewhat by introducing a little newspaper business into it" (Czitrom, 1982, p. 92). With the collaboration of his star student, Robert Park, Dewey tried to start a new kind of newspaper, *Thought News*, to report the latest discoveries of social science and to ameliorate social problems.

Although his utopian newspaper failed, John Dewey never gave up on the mass media's potential for bringing about social reform. Dewey's thinking was based on Darwinian notions of evolution, and on a belief that the newer communication technologies might be able to reconstitute community values in a mass society (Czitrom, 1982, p. 112). Later in his career, John Dewey retreated somewhat from direct involvement in utopian social change, but his interest in newspapers did not cease. John Dewey could be called the first philosopher of communication.

Dewey is known today for his "pragmatic philosophy," a belief that an idea was true if it worked in practice.[1] Pragmatism rejected the dualism of mind and matter, of subject and object. During his decade at the University of Chicago, 1894–1904, Dewey was named to head what became the university's School of Education, including responsibility for a "laboratory school" (an elementary school in which Dewey's philosophical theories were tested). Eventually, Dewey's pedagogical laboratory was considered too radical by the university president, so Dewey resigned from Chicago, and moved to Columbia University (Bulmer, 1984, p. 28).

While he did not conduct experiments himself, Dewey believed that experimentation in particular and science in general provided an instrumental basis for social psychology, a field that Dewey's colleague and friend, George Herbert Mead, was to carry forward in later years at Chicago.

So, in addition to his key role as an intellectual root of communication science, Dewey should be credited with launching the school of pragmatic philosophy, the study of education, and for forming the preconditions for social psychology. Dewey was a direct academic influence on Mead, Park, and Cooley, the other three American roots of communication science.

Charles Horton Cooley: The Looking-Glass Self

The central theoretical interest for Cooley was in how individuals are socialized. Perhaps this scholarly concern grew out of the nature of Cooley's personality. He was acutely shy and suffered from a speech impediment, was introverted, and lead a reclusive life. Cooley (1864–1929) was born in Ann Arbor, Michigan, attended the University of Michigan, and taught there the rest of his life. Initially attracted to the study of sociology by exposure to Herbert Spencer's social Darwinism (it was Dewey who first called Spencer's works to Cooley's attention), Cooley later rejected hereditary and individualism as determinants of personality. Instead, Cooley saw interper-

sonal communication with parents and peers in the individual's
"primary group" as the main basis of socialization.

Cooley placed a high value on communication in his conceptual
schema; it was *the* crucial mechanism in formation of the "looking-
glass self." Interaction with others serves as a kind of mirror, helping
to form an individual's self-conception. For Cooley, communication
provided the means of socialization, and thus represented the thread
that holds society together (Czitrom, 1982, p. 96). The main empirical basis for Cooley's theories came from his own introspection,
and from closely observing how his two small children grew up.

Cooley wrote three main books: *Human Nature and the Social
Order* (1902), *Social Organization* (1909), and *The Social Process*
(1918). The first volume set forth his conception of the looking-glass
self; the two later books dealt more with utopian futures (which
mass communication was to help bring about, as the Ann Arbor
scholar saw it).

Cooley was more focused on interpersonal communication than
on the mass media, although he gave some attention to the latter,
especially in his senior years when his interests turned to utopian
considerations. Cooley had a fundamental belief in social progress,
and perhaps because he was so turned off by the commercialism of
newspapers, he made little effort to explore the realities of mass communication effects, to investigate trends in corporate ownership of
the media, or to determine the role of the mass media in childhood
socialization.

ROBERT E. PARK AND THE CHICAGO SCHOOL OF SOCIOLOGY

Robert Park can rightly be called "the first theorist of mass communication" (Frazier and Gaziano, 1979, p. 1). Perhaps he also
deserves the title of "first mass communication researcher," in that
Park conducted empirical studies of newspaper content, their audiences, and ownership structures. Not the least of Park's accomplishments was his eminence as leader of the Chicago School of
Sociology, one of the most influential centers of social science that
has ever existed.[2] Yet Park did not hold a university position until he
was fifty years old.

Park has been called "perhaps the single most influential person
in American sociology" (Boskoff, 1969, p. 94), while another observer stated that "probably no other man has so deeply influenced

the direction taken by American empirical sociology" (Turner, 1967, p. ix). These are words of very high praise, and Robert Park earned them not as a brilliant lecturer (which he certainly was not), but rather for this tremendous influence on doctoral students at Chicago. Until Park, the great scholars of social science worked in solitude, relatively isolated from other intellectuals with similar concerns. Park changed all this, instituting a "dense, highly integrated local network of teachers and graduate students carrying out a program of research in one city centered around common problems" (Bulmer, 1984, p. 1). Theory and research were blended by Park; he told his graduate students "Go get the seat of your pants dirty in real research" (Bulmer, 1984, pp. 97–98). Robert Park turned social science toward a closer and deeper involvement in the world of which that science was part. The empiricism of communication science, as well as its concern with social change (relatively unique among the American social sciences), trace from the profound intellectual influence of Robert E. Park.

Robert Park was one of John Dewey's most devoted protégés at the University of Michigan; he took six courses from Dewey, and later joined him in the abortive *Thought News*. Dewey implanted a reform-mindedness in his student, which led Park to initially pursue journalistic reform, and later to focus on social science as a means of social amelioration.

Upon his graduation from Michigan in 1887, Park became a journalist, working as a newspaper reporter for eleven years in Minneapolis, Detroit, Chicago, and New York. During this period, he developed a keen ability to observe behavior, particularly the deviant activities (prostitution, delinquency, etc.) of the urban poor (social problems that were later to hold a strong scientific interest for him). He also began to explore how a different kind of journalism might become a powerful tool for social change in America.

His growing interest in the role of news in shaping public opinion eventually led Park to quit his newspaper job in order to enroll for a master's in philosophy at Harvard University. Next, Park went to Germany for his Ph.D., where he studied with Georg Simmel at the University of Berlin. In fact, the three courses that Park took from Simmel were the *only* formal training in sociology he ever received. In 1904, his dissertation, *The Crowd and the Public*, explored the formation of public opinion by the media. It was mainly a theoretical treatise, in the German academic style of that day, but it also contained empirical data to back up Park's theories.

When Park returned from Europe, he did public relations work for the Congo Reform Association and for Booker T. Washington, the great black educator at Tuskeegee Institute. This experience fit with Park's personal and intellectual interest in race relations. Finally, in 1914 at the age of fifty, Park was invited to join the sociology department at the University of Chicago.

The University of Chicago had been founded only in 1892 (about 250 years later than Harvard), with generous gifts from John D. Rockefeller, who had built the Standard Oil empire, and from Chicago industrialists. With Chicago's salaries, which were then about double the prevailing rate for American professors, the new university was able to attract an outstanding faculty. Within a few decades, the University of Chicago was to represent world-class academic quality in the United States. It took the best of new ideas from existing American and European universities and improved on them. Chicago had the first department of sociology in the United States. The university was located on the south side of the city's downtown, in the midst of a lower-class slum inhabited by recent European immigrants, many of whom worked in the nearby meat packing industry. Not surprisingly, given its environment, the Chicago School of Sociology focused on the city of Chicago as a kind of natural laboratory. And Park was the intellectual leader of this ecological approach.

The roots and founders of communication science in the U.S. were *positivists* in several ways; they believed that the methods of science could be applied to the study of society in order to provide useful solutions to social problems. They believed that individuals could harnass scientific methods to improve society. The usual rate of progress, it seemed, could be accelerated by science. Society did not seem to be progressing very much in the early days of the Chicago School (1915–1925): Migration to the United States by the poor of Europe, overurbanization in large cities such as Chicago, and the related social problems of crime and prostitution characterized the nation. The Chicago School's emphasis upon urban social problems is illustrated by the titles of the early books it produced: *The Hobo* (1923), *The Gang* (1927), *The Ghetto* (1928), *The Gold Coast and the Slum* (1929), *The Negro Family in Chicago* (1932), *Vice in Chicago* (1933), and *10,000 Homeless Men* (1936).

The Chicago School of Sociology attacked the "instinct" theory of human behavior, arguing instead that an individual's behavior could be explained by the interactionist social psychology propounded by John Dewey, Charles Horton Cooley, and George Herbert Mead. The biological determinism of instincts was attacked theoretically, and instincts were shown to lack an empirical reality. According to the interactionists, a child is born neither human nor social, but soon learns language and a sense of self, which is what makes the individual a social being. This view of personality development put interpersonal communication, especially that with the primary group of parents, peers, and friends, in a central place. The self, which Cooley picturesquely called the looking-glass self, is attained when a child becomes able to imagine how he or she appears to others, how others evaluate his or her behavior, and when the child reacts to this evaluation (Hinkle and Hinkle, 1954, pp. 30–31). So instead of instincts as an explanation for human behavior, the Chicago sociologists showed that behavior, especially the deviant behavior that they studied, was group-influenced by means of communication via interpersonal networks. For example, a Chicago professor told his classes that he had never met a boy who committed his first delinquent act alone (Faris, 1970, p. 76).

Thus, communication was a fundamental human process to the Chicago sociologists, although other than Robert Park, they did not point the specific directions that future communication research was to take. Perhaps this is one reason why the relationship of the Chicago School of Sociology to the modern discipline of communication science has been so little appreciated, and most often ignored by other observers tracing the history of communication research.

Park defined communication as "a social-psychological process by which one individual is able to assume, in some sense and to some degree, the attitudes and the point of view of another." This subjectivistic explication avoids the linear, one-way thinking later implied by the communication models based upon the information theory of Claude Shannon and Warren Weaver in the 1950s. Park's conception of communication allowed that two or more individuals might exchange information during the communication process, with each giving a different meaning to the information they received.

Robert Park not only theorized about communication, he also investigated mass communication in several ways that are now utilized by contemporary scholars. For instance, Park published a research monograph entitled *The Immigrant Press and Its Control* (1922), which dealt with the role of foreign-language newspapers in

America. Park showed that Yiddish, Polish, German, and other newspapers for immigrants actually slowed down their acculturation into mainstream American life.

The notion of agenda-setting, through which the mass media establish the priorities of news issues and thus influence what audience members talk about, rests on Park's identification of news as a basis for interpersonal discussion (Frazier and Gaziano, 1979, p. 2). Park thought of public opinion as measurable, and thus was far ahead of George Gallup, Paul F. Lazarsfeld, and other survey pioneers of the 1940s. Robert Park was also advanced for his time in raising such research questions as: How are interpersonal networks linked to the mass media? To what extent does the newspaper influence public opinion (this is agenda-setting) and how is the newspaper controlled by public opinion? How capable are the media in bringing about social change?

After Robert Park, American sociology turned from communication to the study of work and social class, and communication has not been a central issue of sociological inquiry for over fifty years. And after a twenty-year heyday in dominating the field of sociology (until 1935), the Chicago School declined to become just another excellent sociology department.

GEORGE HERBERT MEAD: THE SELF

Mead (1863–1931) studied pragmatic philosophy with William James at Harvard and did graduate work in Germany, but he was most influenced by his colleague at the University of Michigan, John Dewey. Mead moved to the Chicago Department of Philosophy in 1894 at the invitation of Dewey, and became involved in various social action projects, after forty previous years of purely philosophical work. Mead directed a survey aimed at reforming the life of Chicago's stockyard workers, and also was involved in Dewey's laboratory school (Bulmer, 1984, p. 24). In fact, Mead soon became known as one of Chicago's "leading reform figures" (Bulmer, 1984, p. 124). Such progressivism was closely related to Mead's *behaviorism*, an academic belief in the study of the individual's experience from the point of view of conduct, especially as it is observed by others.

Mead's ameliorative orientations fit with the Chicago School of Sociology, in which Mead was an intellectual influence second only to Robert Park (even though Mead was a professor in the Department of Philosophy, all of the doctoral students in sociology took his course in Advanced Social Psychology). During his thirty-seven years of teaching at Chicago, Mead founded the symbolic interactionist approach to social psychology. Strangely, Mead was influential not by writing books, but through his teaching. In fact, he never published a book. A few years after he died, Mead's students, drawing from their classroom notes, published *Mind, Self and Society* (Mead, 1934), the core volume about symbolic interactionism. Mead depended directly upon his friends John Dewey and Charles Horton Cooley for his theory of the self, but carried their work forward in a direction stressing human communication as the fundamental socializing agency. Mead's theory states that individuals get to know themselves through interaction with others, who communicate to them who they are.

Mead emphasized that the self begins to develop in a child when the individual learns "to take the role of others," to learn to imagine the roles of others, and to anticipate their responses to the individual's actions. This empathic ability rests on the use of language and is learned by social interaction with one's primary groups. So the interactionist social psychologists recognized communication as a basic human process. Mead (1934, p. xxiv) created the concept of the "generalized other," with whom one learns to empathize. Thus the "me" consists of all the attitudes of other individuals with whom one has interacted, and which one takes over into oneself. "Me" is an individual's perspective of how others see him or her.

What did the four American "roots" of communication science have in common? They studied, and taught, at high-prestige American universities, Michigan and Chicago, and, in the case of Park and Mead, studied in Germany. Thus they bridged from European to American scholarship. Their institutional affiliation gave power to their ideas about human communication. All four "roots" came from rural backgrounds, but led urban lives, and this transposition influenced their progressive liberalism. Dewey, Cooley, Park, and Mead were positivists, believing in the ameliorization of social problems through social science research. American newspapers had become important mass media, reaching huge audiences by 1900; the four progressive scholars were fascinated with the potential of newspapers for social change. All four were empiricists in their way, using data-gathering methods ranging from Cooley's

introspective observations to Park's content analyses and Mead's encounter with survey studies. All were humanistic philosophers in
their orientation, emphasizing theory-development but also stressing the role of "getting one's pants dirty" with empirical data. The
four were before their time in recognizing communication as the
fundamental process affecting human behavior. They stressed the
subjectivity of human communication, a quality that was later to be
unfortunately short-changed by linear, effects-oriented models of
communication. The four studied a very wide range of communication phenomena, pursuing a diversity of research questions not seen
since. The paradigm that was to unify and integrate the communication discipline would not come until the next era of communication research.

The Engineers of Communication: Claude Shannon and Norbert Wiener

The several years just after World War II marked a particularly
crucial time for development of the new field of communication
research. It was as if these important technological and social ideas
had been dammed up by America's wholehearted dedication to
fighting the war. Interdisciplinary research had flowered in various
military projects during the war, and this cross-fertilization was important to the emerging communication science.

Within a few years of 1945, several important events happened
in quick succession that were to advance the technological underpin-
nings of human communication.

• In January 1946, the first mainframe computer, ENIAC, was
switched on at the Moore School of Engineering at the University of
Pennsylvania. Thus, the computer revolution was launched, although Penn was soon to lose its dominant position in computer
technology when university officials tried to force the creators of
ENIAC, Presper Eckert and John Mauchly, to sign over their commercial rights.

• A hundred miles or so from Philadelphia (at Bell Labs in Murray Hill, New Jersey) and a few years later, William Shockley, John
Bardeen, and Walter Brattain invented the transistor, probably the
most important discovery of the century. The transistor was a solid-
state device that was to replace the vacuum tube as the fundamental
unit of electronics. The advantages of the transistor are its small size

and its low requirement for electrical power. By the 1980s, semiconductor chips (a modern-day kind of transistor) were to miniaturize computers, making them so low-priced that most American homes and businesses could afford them.

• The late 1940s were also years of important scholarly work in communication science. At MIT, the mathematician Norbert Wiener published his *Cybernetics, Or Control and Communication in the Animal and the Machine* (1948). Two years later, Wiener wrote a nontechnical companion volume that also became a bestseller called *The Human Use of Human Beings: Cybernetics and Society* (1950).

• Even more influential in the rise of communication science was a small book by Claude E. Shannon, an electrical engineer at Bell Labs, and Warren Weaver of the Rockefeller Foundation in New York. *The Mathematical Theory of Communication* (1949) proposed a general model of communication accompanied by a series of mathematical theorems about the engineering aspects of communication. This simple model was accepted with enthusiasm by communication scholars, and was to influence their research for the decades that followed. Unfortunately, these various scholars distorted the original model in certain very fundamental ways (as we show later).

• About this same time, Wilbur Schramm, an academic convert from the field of fiction writing, moved from the University of Iowa to the University of Illinois, where he launched the first Ph.D.-granting school of communication. Many scholars in previous years had conducted communication studies, but they identified mainly with their parent disciplines of sociology, psychology, or political science. They pioneered in communication research, but it was then an intellectual territory "where many ventured, but few tarried" (Schramm, 1959). After Schramm at Illinois, communication research became institutionalized in university institutes, schools, and departments of communication. Thereafter, many ventured and most stayed in the new discipline.[3]

THE MATHEMATICAL THEORY OF COMMUNICATION

Shannon and Weaver's *The Mathematical Theory of Communication* (1949) was published at the time when "Communication researchers were emerging from the ghetto of trade-school education, and straining for the intellectual respect of their academic colleagues" (Berlo, 1977, p. 16). Information theory offered a means by which communication scholars could gain academic respectability. What the communication scientists adopted, however, was a somewhat mistaken conception of Warren Weaver's communication model. They did not test Claude Shannon's theories of information, nor, with very few exceptions, use his entropic measure of the amount of information.

"Information theory was an engineer's discovery" (Campbell, 1982, p. 16). This theory was first presented to the world in two papers by Claude E. Shannon published in the July and October 1948 issues of the *Bell System Technical Journal*. Shannon proposed a set of theorems, expressed in mathematical form, dealing with the sending of messages from one place to another. "But the wider and more exciting implications of Shannon's work lay in the fact that he had been able to make the concept of information so logical and precise that it could be placed in a formal framework of ideas" (Campbell, 1982, p. 17). Shannon treated information as a general concept, dealing with all kinds of information, everywhere.

Jeremy Campbell, an unabashed admirer of Shannon's theory, stated: "Information emerged as a universal principle at work in the world, giving shape to the shapeless, specifying the peculiar character of living forms and even helping to determine, by means of special codes, the patterns of human thought. In this way, information spans the disparate fields of space-age computers and classical physics, molecular biology and human communication, the evolution of language and the evolution of man" (Campbell, 1982, p. 16). Shannon's information theory was published only five years before James Watson and Francis Crick discovered that the structure of DNA, the double helix, was an information system (Campbell, 1982, p. 19). Key scientific discoveries such as Einstein's relativity, Noam Chomsky's linguistics, and brain functioning can best be understood in light of Shannon's information theory (Campbell, 1982). After Shannon, "Nature must be interpreted as matter, energy, and information" (Campbell, 1982, p. 16). Indeed, the impacts of information theory immediately following the publication of the Shannon and Weaver book occurred in disciplines ranging from electrical engineering to biology to brain research (Dahling, 1962).

The timing of Shannon's theory was crucially related to electronics, mass communication, and computers. "Information became a scientific concept when the age of electronic communication dawned" (Campbell, 1982, p. 16). The key advances being made in the electronic technologies of human communication opened the way for the rapid adoption of the Shannon paradigm. And at the heart of Shannon's theory was the concept of information.

What is *information?* Patterned matter-energy that affects the probabilities of alternatives available to an individual making a decision. Claude Shannon measured information in terms of the choices available to a source in the process of forming a message (we shall discuss his entropic measure of information shortly).

Claude E. Shannon (1916–   ) was born in the small city of Petrosky, Michigan, earned his bachelor's degree at the University of Michigan, and went on to MIT for his Ph.D. He spent his career at Bell Labs until 1956, when he moved back to an endowed chair at MIT. Shannon was "famously unprolific" (Campbell, 1982, p. 20) in that his perfectionist work style seldom allowed him to finish a publication. He simply could not bear to surrender a paper that he had written. Shannon avoided recognition for his accomplishments, he did not answer letters, he did not like to teach, and eventually he just stopped coming to his office at MIT, preferring instead to play the stock market (where he could test the usefulness of his information theory to his heart's content).

Shannon's information theory was influenced by various academic ancestors, several of which Shannon (1949, pp. 31–32, 85) acknowledged: H. Nyquist (1924) and R. V. L. Hartley (1928) for providing the basis of a general theory of communication; J. W. Tukey for the idea of using "bit" as the unit of information; Norbert Wiener's (1949) *Cybernetics* for "the first clear-cut formulation of communication theory as a statistical problem." Not cited was R. A. Fisher, the great English statistician, who, around 1915, had formulated a concept of information similar to Shannon's.

SHANNON'S INFORMATION THEORY

Without doubt, the Shannon and Weaver (1949) paradigm[4] was the most important single turning point in the history of communication science. That fact was immediately obvious to Wilbur Schramm at the University of Illinois, where the first academic program in communication science was getting underway. Louis Ridenour, a physicist, was dean of the graduate school at Illinois. He realized the intellectual power of Shannon's two journal articles when they appeared in 1948, showed them to Schramm, who was editor of the University of Illinois Press (among his other activities at Illinois), and suggested that they invite Warren Weaver of the Rockefeller Foundation to write a popular introduction (Dahling, 1962).[5] The University of Illinois Press published the two chapters as a small book. Within four years, *The Mathematical Theory of Communication* sold 6,000 copies in hardback; by 1985, over 32,000 paperback copies had been sold,[6] and the book continues to sell well.

Weaver's chapter is primarily speculation about how a human communication theory might be developed out of Shannon's mathematical theorems about engineering communication. Warren Weaver, as an engineer without a background in behavioral science, might not have been the most appropriate person to fill this translator role for Shannon's theories. All that Shannon claimed he had contributed was a mathematical model of signal transmission. It was Weaver who extended the model to *human* communication. Weaver also wrote much more readably, without complicated mathematical notations (in contrast, the Shannon chapter in their book is exactly the same as his two articles in the *Bell System Technical Journal*, except for the title, which was changed from "A Mathematical Theory . . ." to "The Mathematical Theory . . ."). Most social scientists of the day did not work their way through Shannon's math, and so they had to be satisfied with Weaver's simplified "pony." Few communication scholars understand that Weaver's essay is not a précis of Shannon's theory, but rather an original work, going on from Shannon's conception of radio transmission into speculation about the extension of information theory to human communication (Ritchie, 1986). Later events showed this misunderstanding to have unfortunate consequences regarding the impact of information theory on research.

Because of the important differences in content between the Weaver and the Shannon chapters in *The Mathematical Theory of Communication*, from here on I shall cite either Weaver or Shannon, unless I am referring to the entire book.

The Shannon and Weaver model of communication (Figure 3–2) led communication scientists into a linear, effects-oriented approach to human communication in the decades following 1949. Shannon's information theory and Weaver's communication model are two separate intellectual contributions. One can adopt one and not the

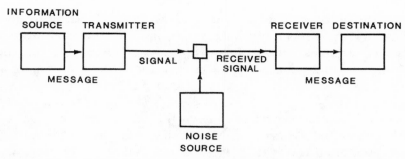

Figure 3-2. The Shannon and Weaver Model of Communication

The components in this model are defined as: "The *information source* selects a desired *message* out of a set of possible messages. . . . The *transmitter* changes the *message* into the *signal* which is actually sent over the *communication channel* from the transmitter to the *receiver*. . . . The *receiver* is a sort of inverse transmitter, changing the transmitted signal back into a message, and handing this message on to the destination. . . . In the process of being transmitted, it is unfortunately characteristic that certain things are added to the signal which were not intended by the information source. . . . All of these changes in the transmitted signal are called *noise*" (pp. 7–8).

SOURCE: Claude E. Shannon and Warren Weaver (eds.), *The Mathematical Theory of Communication* (Urbana: University of Illinois, 1949), p. 34. Copyright 1949 by The Board of Trustees of the University of Illinois. Used by permission.

other. Further, the communication model, on close reading, turns out to be (1) not so linear in nature, and (2) not so antagonistic to the subjectivity-of-communication viewpoint of Dewey, Cooley, Park, and Mead, as most critics of Shannon's model have claimed.

Notice in Figure 3–2 that the model recognizes that a "message" must be encoded into a "signal" by the source, and later decoded from "signal" to "message" by the receiver. Only the signal is sent from the transmitter to the receiver. As Seth Finn and Donald Roberts pointed out: "Shannon's conceptualization of the communication process may be better symbolized by two intersecting circles in a Venn diagram than by the string of components that has come to signify information theory [they mean the communication model]" (Finn and Roberts, 1984). As Weaver emphasized: "*Information* must not be confused with *meaning*" (Weaver, 1949, p. 8). Such a distinction helps put the subjectivity of human communication into Shannon's model of communication. Certainly, Shannon and Weaver realized that the encoding of messages into signals, and their later decoding, was a subjective process when human beings were involved.

Actually, Shannon, Weaver, and Norbert Wiener were more astute about the social psychology of communication than were the mass communication scholars who later followed the Shannon model of communication. The later scholars dropped Shannon's distinction between message and signal in their communication models; an example is David K. Berlo's (1960) S-M-C-R (source-message-channel-receiver), although the subjectivity of communication was often stressed in the continuing tradition of communication research. However, it was more talked-about than studied. Thousands of students in introductory communication courses were taught the S-M-C-R model, a necessary simplification of Shannon's model.

As David Ritchie pointed out: "To criticize Shannon's model as inapplicable to the complexities of human communication is to criticize a rowboat because it is not a whale" (Ritchie, 1986). Later-day communication scholars basically misunderstood the Shannon rowboat because they never looked at it carefully enough.

An immense benefit of Shannon's theory for the field of communication science was to encourage the feeling of universality of communication across the different types of channels. It did not really matter, information theory argued, whether communication occurred via machine or by face-to-face channels. One impact of information theory was to emphasize the basic similarity of interpersonal and mass media communication, which had grown up out of different trade-school traditions in the United States. Shannon implied that communication was communication, through whatever channels it occurred. The net effect of such theoretical thinking was to standardize the terminology, concepts, and model of human communication, whatever the channel.[7]

Perhaps the most important negative impact of the Shannon paradigm was that it headed communication scientists into the intellectual cul-de-sac of focusing mainly upon the *effects* of communication, especially mass communication. Communication is purposive and directional, according to Weaver, who defined communication as "all of the procedures by which one mind may effect another" (Weaver, 1949, p. 3). Shannon (1949, p. 31) stated: "The fundamental problem of communication is that of reproducing at one point either exactly or approximately a message selected at another point" (Shannon, 1949, p. 31). Weaver said that the effectiveness of communication rests on "the success with which the meaning conveyed to the receiver leads to the desired conduct on his part" (Weaver, 1949, p. 5). These quotations equate communication with a linear transmission of messages and with source-control of the receiver.[8]

The concept of feedback was not mentioned per se by Shannon and Weaver (1949), although they cited Norbert Wiener's book, *Cybernetics* (1948). Shannon discussed the idea of a "correcting device" as an addition to his basic communication model (Shannon, 1949, p. 68). So the clues were there for communication scientists to pick up, but they didn't do it. Why not? Because a linear model of communication more conveniently fit the one-way nature of the mass media (newspapers, radio, and television) that communication scholars were studying at the time. So the feedback and dynamic aspects of the Shannon communication model were deemphasized, ignored, and eventually dropped. Here we see how the nature of communication technology influences the direction of communication theory and research.

Once a paradigm is accepted in an academic discipline, it provides useful guidelines for future generations of scholars, removing much of the uncertainty about what to study, how to study it, and how to interpret the ensuing research results. A paradigm is also an intellectual trap, enmeshing the scientists who inherit it in a web of assumptions that they often do not recognize. David K. Berlo, an early and influential communication scholar, on looking back at his Ph.D. training at Illinois in the 1950s, stated: "Like many of my colleagues, I simply did not understand the underlying assumptions and theoretical consequences of what I believed, and had not grasped the limited fertility of the research tradition in which I had been trained. I did not recognize that the assumptions underlying linear causal determinism may account for the major proportion of communication events, but not account for the portion that makes a significant difference in our lives" (Berlo, 1977, p. 12). This admission is especially important because Berlo's book, *The Process of Communication* (1960), was the most important means of popularizing Shannon's communication model to students of human communication.

Several decades after publishing his more-or-less linear-effects model of communication, based on Shannon's model, Berlo recognized that human communication is often nonlinear: "An information-communication relationsip may be directional as we conceive it, or it may not. If we look at the 'source' as intentional and initiatory and the 'receiver' as passive and a receptive container—

e.g., if the message is stimulus and the effect is response—the relationship is directional. On the other hand, if the relationship is one in which both users approach the engagement with expectations, plans, and anticipations, the uncertainty reduction attributable to the contact may better be understood in terms of how both parties use and approach a message-event than in terms of how one person uses the contact to direct the other (Berlo, 1977, p. 20).

So, unfortunately, what communication scholars interpreted of the Shannon paradigm was structured by their needs and interests of the day. What they wanted was a linear, one-to-many model of human communication. And that's what they got.

THE IMPACT OF SHANNON'S THEORY

Shannon's model could not really be tested by scholars of human communication. It had too many component parts, linked in too many complex ways. So the model mainly served a sensitizing function, alerting scholars to dimensions of communication that they otherwise might not have appreciated. Shannon's theorems, derived from his model, could have been tested by communication scientists. This, however, they did not do, mainly because they could not understand them.

Claude Shannon also suggested a measure for the concept of information, defined as patterned matter-energy that affects the probabilities of alternatives available to an individual making a decision:

$$H = - \Sigma p_i \log_2 p_i = p \log_2 (1/p)$$

This equation for the amount of information was the same equation as that devised by nineteenth-century physicists for the entropy principle. *Entropy* is the degree of uncertainty or disorganization of a system. Claude Shannon suggested that the amount of information could be measured by the logarithm of the number of available choices, with the logarithm calculated to the base 2, rather than to the more usual base 10. Thus, the unit of information is a "bit" (a word first suggested by the statistician John W. Tukey as a condensation of "binary digit"). Each of the two alternatives should be equally probable for the choice to represent one bit of information. The choice by a source of one of sixteen alternative messages, among which the source is equally free to chose, represents four bits of information ($16 = 2^4$).

Shannon's entropy measure has been utilized by several communication scholars to measure: the selection of daily news from various sources by *The New York Times* (Schramm, 1955); the variety of news issues placed on the public agenda in a community by the mass media (Chaffee and Wilson, 1977); to measure the readability of print messages (Taylor, 1953 and 1956); and to operationalize the contraceptive behavior of Korean villagers as the result of an information campaign (Rogers and Kincaid, 1981, p. 283). The uses of Shannon's entropy measure of information are very rare. The entropy measure deserves more attention from communication scientists than it got (Finn and Roberts, 1984).

Finally, Shannon's communication model failed to create a unified model of human communication because "the theory did not consider the semantic or pragmatic levels of communication" (Rogers and Kincaid, 1981, p. 33). Electronic communication via radio, television, and film usually *is* linear. One cannot really fault Shannon for seeming to propose a linear model (after all, he was pointing out that the bidirectional telephone was really, in essence, two unidirectional transmission systems). But the communication scientists who followed the Shannon model in their research and in their theory-building also assumed a linearity of human communication. "These models described a simple communication act, but not the *process* of communication. Many important aspects of human communication do not fit linear models, and tended to be ignored by communication research based on linear models" (Rogers and Kincaid, 1981, p. 34).[9]

NORBERT WIENER AND CYBERNETICS

The other great "engineer" of communication theory was Norbert Wiener, the father of cybernetics.[10] Wiener and Claude Shannon both made MIT their academic headquarters, with Wiener spending his entire forty-five-year career in MIT's Mathematics Department, where he served until his death in 1964. He taught Shannon in the early 1930s, and Wiener and Shannon both worked on military research projects at MIT during World War II. They were not intellectual collaborators, although their contributions to communication theory bear a certain relationship to each other. As Wiener put it in his book, *I Am a Mathematician:* "Shannon and I had relatively little contact during his stay here [at MIT] as a student. Since then, the two of us have developed along parallel if different directions, and our scientific relations have greatly broadened and deepened" (Wiener, 1956, p. 179).

As Francis Bello, put it: "To Norbert Wiener goes the major credit for discovering the new continent and grasping its dimensions; to Claude Shannon of Bell Laboratories goes the credit for mapping the new territory in detail and charting some breath-taking peaks" (Bello, 1953, p. 137). Although Norbert Wiener's theory was, for several years, only available in an obscure military technical report on antiaircraft control, he deserves credit for co-inventing the entropy measure of information usually associated with Claude Shannon's name; Wiener (1956) refers to "the Shannon-Wiener definition of quantity of information (for it belongs to the two of us equally)."

The main difference in the Shannon versus the Wiener approach to communication theory was that Shannon dealt mainly with discrete (yes/no) messages, while Wiener's theory concerned continuous flows of information. "I approached information theory from the point of departure of the electrical circuit carrying a continuous current" (Wiener, 1956, p. 263). Such an approach was fundamental to cybernetics, the science of maintaining order in a system, which takes its name from the Greek word meaning "steersman." All systems have a tendency to become entropic or disorderly bcause of their random deviations from order that must continually be corrected (Campbell, 1982, p. 23). Information about the system's performance is fed back to keep the system on course.

Norbert Wiener was one of the finest mathematicians the United States has produced (Campbell, 1982, p. 23). He was the son of a Harvard professor of Slavic languages, who expected superhuman intellectual accomplishments from his son. He got them. Norbert Wiener was a university student at age ten, and had completed his Ph.D. dissertation at Harvard on the relationship between mathematics and philosophy by age eighteen. Then he studied under Bertrand Russell at Cambridge University as a post-doc, before moving on to further post-doctoral work in Germany at the University of

Gottingen. Appropriately, Norbert Wiener's autobiography is entitled *Ex-Prodigy* (1964b).

In his appearance and manner, Norbert Wiener exemplified the stereotype of the university professor, almost as if he had been sent over from Central Casting to play the role. He was a short, stout man with a paunch, wore a small white goatee, and tended to walk splay-footed in a ducklike fashion, usually with a cigar in hand. Wiener was an unstoppable talker (Heims, 1980, p. 206). During academic conferences, Wiener had a famous tendency to go to sleep and even to snore while others were talking; immediately upon awakening he would often make a penetrating comment, showing that he understood all that had been said.

THE YELLOW PERIL

Wiener's cybernetic theory grew out of his wartime research on improving the accuracy of antiaircraft fire, and from Wiener's intellectual exchanges with another brilliant Jewish mathematician, John von Neumann at the Institute for Advanced Study at Princeton University. Von Neumann had provided certain of the key ideas for the invention of ENIAC. He realized that a computer essentially performed logical functions, and that the electronic aspects were really secondary (Heims, 1980, p. 182). Von Neumann was a consultant to the Manhattan Project, making frequent trips to Los Alamos, New Mexico, during World War II. He is credited with carrying out the complex mathematical work on computers that led to the implosion detonating device for the first atomic bomb.

The friendship between von Neumann and Wiener was eventually destroyed by jealousy; von Neumann felt that he had not been given enough credit by Wiener for his thinking about the concept of feedback, and for drawing the connection between entropy and information (Heims, 1980, p. 208).

Wiener's research project during World War II in which he developed cybernetic theory concerned the problem of antiaircraft fire control. This work was funded by the U.S. military and was carried out at MIT under the general direction of Warren Weaver, then an official in the U.S. Defense Department. The scientific problem of improving gunnery accuracy was given the highest national priority. At that time, existing antiaircraft guns were almost powerless against German bombers—a fact proven in 1940, during the Battle for Britain, when 10,000 shells were fired for every German plane destroyed.

Wiener approached the gun-control research as a problem of information theory. Each shell fired was a message; how closely it approached the target was then information to be fed back to the electronic gunsight of the antiaircraft weapon, so that the next shot (a few seconds later) would be more accurate. The imperfect state of radar at that time meant that unwanted electrical interference, or "noise," was present in the communication system. After a frenetic research effort, with Wiener popping benzedrine pills to keep awake during marathon work sessions, he completed a classified report that was circulated to fellow scientists. Because of the color of its cover and due to the intimidating difficulty in understanding its contents, this monograph came to be known as the Yellow Peril (Campbell, 1982, p. 30).

The Yellow Peril was a major contribution to communication theory, launching cybernetic theory, although Wiener's ideas did not become publicly known until several years later with the publication of his best-selling books on the topic. Whether Wiener's research directly improved the accuracy of antiaircraft fire is not entirely clear, although the cybernetics-designed gunsight equipment brought about a startling improvement in accuracy. In August 1944, before this technology was used in defending England from German buzz-bombs, only 10 percent were shot down. Afterward, 50 percent were destroyed (Campbell, 1982, p. 31).

THE IMPACT OF WIENER'S CYBERNETIC THEORY

Although Norbert Wiener was a mathematician at an engineering school, his intellectual interests ranged widely to include physics, neural physiology, and psychiatry. He collaborated with scientists specializing in these and other fields, often co-authoring books and articles.

An informal study group of about twenty key scientists met regularly with Norbert Wiener over the period from 1945 to 1952 in order to explore the applications of cybernetic thinking to their fields. These Conferences on Cybernetics were sponsored by the Macy Foundation and usually held in a New York hotel. Wiener was the star. Among the invitees were Kurt Lewin, George Herbert

Mead, and the other social scientists who played heavy roles in the founding of communication science (Heims, 1980, pp. 199–210). Von Neumann was also an important contributor at these sessions, until his break with Wiener.

During this same post-war period, Wiener's best-selling books, *Cybernetics* (1948) and *The Human Use of Human Beings* (1950), were published. Wiener became a widely known personality, described by news magazines as a Santa Claus figure with white beard and round tummy. Wiener said that he published these two books in order to write himself out of a financial hole, as an alternative to accepting military contracts (Heims, 1980, p. 335). In peacetime, Wiener felt it was wrong to conduct military research, in which the powerful new technologies would be used for destructive purposes.

Given the popularity of his ideas among the educated public, and the interest of social scientists in his cybernetic theory, it might seem puzzling that communication science was not affected more directly by Norbert Wiener. There are several possible reasons:

1. Wiener refused to become involved in extending cybernetic theory to human communication because he felt that the problem was so much more complex than with machine communication (Heims, 1977). He even warned the social scientists of his day against seeking to apply cybernetics to human systems.

2. The competing theory of Shannon and Weaver, much simpler to understand (or so communication scholars then thought), seemed to better fit the research interests of communication scientists in 1950. They wanted to investigate the direct effects of such one-way mass media as television; a linear communication model was perfect for this purpose, and a cybernetic model with feedback loops and continuous flows was not. Further, a prestigious figure in the communication field, Wilbur Schramm, was behind the Shannon/Weaver theory, or so it then seemed.[11]

So for many years communication research was to focus on a linear effects model of human communication. Finally, in the 1980s, a veteran communication scholar, Elihu Katz, was able to note that "The best thing that has happened to communication research is that it has stopped frantically searching for evidence of the ability of the media to change opinions, attitudes, and actions in the short run" (Katz, 1983). Perhaps communication science could have saved twenty to thirty years of misdirected activity if its scholars had attended more carefully to the information theory set forth by the engineers of communication.

Our present discussion of the engineers of communication illustrates how the 1980s interest of communication scientists in the
new technologies have forced us to redefine who our founders are.
Today we see increased attention being given to Claude Shannon
and, especially, to Norbert Wiener's cybernetics.

The Four Founders:
Lasswell, Lewin, Hovland, Lazarsfeld

Four key intellectuals came from their own disciplines into communication research and made such impacts that they should be
regarded as the fathers of the field (Schramm, 1985, p. 201). The
four are Harold Lasswell (1902–1980), the political scientist; Kurt
Lewin (1890–1947), the social psychologist; Carl Hovland (1921–
1961), the experimental psychologist; and Paul F. Lazarsfeld (1901–
1976), the sociologist.

They had strikingly similar careers. "All of them had [intellec-
tually] rich early backgrounds, went to excellent universities, came
into contact with great minds. All of them were broadly interdisciplinary by inclination. All of them went through abrupt career
changes in midlife, and turned from their own discipline to communication through the experience of confronting 'real world' problems. All of them gathered around them younger scholars who came
to be leaders in the field" (Schramm, 1985, pp. 201–202). In their
communication research, all four founders focused on *effects*, the
changes in an individual's behavior (knowledge, attitudes, or actions) that occur as the result of the transmission of a communication
message. The exact effects that each founder studied differed, of
course, with Harold Lasswell investigating the impacts of propaganda on public opinion and Carl Hovland experimenting with the
consequences of source credibility on individuals' attitude change.

The effects-orientation of the four founders fit well with the ag-
gressively empirical spirit of American social science that was growing stronger in the 1930s. Improved sampling techniques, more accurate means for measuring attitudes, and other increasingly
sophisticated research methods facilitated the move to greater empiricism. "Emphasis on the evaluation of the scientifically measurable behavioral effects of modern communication provided intellectual pollen for the cross-fertilization of methodologies, disciplines,
and institutions in communication research" (Czitrom, 1982, p.

122). The four fathers of communication research converted the philosophical and theoretical orientation of the emerging field to one that became very heavily empirical. Thus the abstract ideas of Dewey, Cooley, Park, and Mead, and the communication models of Shannon and Wiener, were replaced by the evidence-supported generalizations about communication effects found by Lasswell, Lewin, Hovland, and Lazarsfeld. Communication science would never be the same again.

The effects-orientation of the four founders was in tune with their times. For example, Harold Lasswell investigated the effects of propaganda messages that had been used in World War I; after the 1919 Armistice, there was great public interest in this issue. In his *Propaganda Technique in the World War*, Lasswell (1927) analyzed the main strategies used during the war and their effects.[12] In the early 1930s, the Payne Fund studies (Charters, 1934) dealt with the effects of movies on children; a typical question, asked of child-respondents in the Payne Fund studies, was "Do you have bad dreams after seeing a movie?" As this question implies, the focus was upon identifying the negative effects of film on children. Finally, radio broadcasting was becoming important in America during the 1930s, and so when Paul F. Lazarsfeld arrived from Vienna, he was soon offered the directorship of Princeton University's Office of Radio Research in 1937. Two years later, he moved with this institute to Columbia University, where it became the Bureau of Applied Social Research. This change in name was socially significant in that Lazarsfeld saw radio research as a direct step toward a more general type of communication research (Czitrom, 1982, p. 130).

Unfortunately, there was an intellectual cost to the effects-orientation of communication research. Other important issues could not be studied, thanks to the overwhelming emphasis upon studying effects. For example, questions of ownership and control of the mass media in America tended to be ignored; these were research topics that earlier scholars—for example, Robert Park—had regarded as crucial, as do the "critical school" of communication scholars today.[13] Critical scholars have a valid point, I think, when they crack down hard on empirical communication scholars for ignoring who owns and controls the mass media. Study of the context of communication was shortchanged by the heavy emphasis upon effects.

Now we take up each of the four founders and their contributions, describing in more detail the synthesis shown in Table 3–1.

Table 3–1. The Four Founding Fathers of Communication Science

Founder/ Discipline	Research Methods Used	Typical Independent Variables/ Dependent Variables	Main Books
1. Harold D. Lasswell/ Political science	Content analysis	Symbols of identification, image versus reality/ *effects* on public opinion	*Propaganda Technique in the World War* (1927)
2. Kurt Lewin/ Social psychologist	Group experiments in quasi-natural settings	Gatekeeping, autocratic versus democratic leadership style, group pressures on individuals/*effects* on individual behavior of group members	*Informal Social Communication* (written by Lewin's students); *Principles of Topological Psychology* (1936)
3. Carl Hovland/ Experimental psychologist	Laboratory experiments	Source credibility, one-sided versus two-sided messages, fear appeals/ *effects* on persuasion (attitude change)	*Experiments in Mass Communication* (with others) (1949); *Communication and Persuasion* (with others) (1953)
4. Paul F. Lazarsfeld/ Sociologist	Personal interviews, surveys	Socioeconomic status, mass media exposure, interpersonal influence/ *effects* on individual's knowledge, attitude, and behavior change	*The People's Choice* (with others) (1944); *Communication Research, 1948–1949* (with Frank Stanton) (1949); *Personal Influence* (with Katz) (1955)

SOURCE: Based somewhat on Berelson (1959).

HAROLD D. LASSWELL: PROPAGANDA EFFECTS

To identify Harold Lasswell as a political scientist is not entirely accurate. His was such a wide-ranging mind, and so original, that he does not fit neatly into a disciplinary box. For instance, Lasswell was the first American social scientist to become interested in psychoanalysis, and he studied with Theodore Reik in Berlin in 1929–1930, where he underwent therapy. Several of his writings, such as *World Politics and Personal Insecurity* (Lasswell, 1953), show a strong Freudian influence. As one of Lasswell's Chicago colleagues remarked: "[The department head] sent him to England and he came back with an English accent, he sent him to Vienna and he came back with a full-grown psychoanalytic vocabulary, and he sent him to the Soviet Union and when he came back he showed that Marx could be reconciled with Freud" (Harold Gosnell, quoted in Bulmer, 1984, p. 194).

Lasswell was such an enthusiastic learner, he became genuinely interested in every possible problem that he encountered in daily life. For instance, his interest in Freudian psychiatry arose one summer when he was visiting his uncle, a medical doctor in Indiana, who was unable to cure a patient with a paralyzed arm. Lasswell's uncle ordered a set of Freud's books sent from Vienna, which the young Harold read. They seemed rather sensible. "It was not until I was a junior at the University of Chicago that I discovered that Freud was controversial," said Lasswell (Reston, 1969, p. 6). Later, Lasswell was to be the leading U.S. scholar in introducing Freudian theory into political science, and, more broadly, into American social science. Lasswell was very broad-gauge: "By the time he was well into graduate school, Lasswell was publishing across the board in the social sciences" (Smith, 1969, p. 51). His articles and books dealt with economics, sociology, political science, among other fields; he has been described as "a kind of Leonardo da Vinci of the behavioral sciences" by one of his biographers (Smith, 1969, p. 41). Such an interdisciplinary mind could not ignore such a fundamental human process as communication. In fact, what we would today call a communication point of view pervaded much of Lasswell's thinking and writing, whatever the exact topic of scholarly concern.

Harold D. Lasswell was born in Donnellson, Illinois, population 292. He was a precocious child, and by age sixteen was enrolled at the University of Chicago on a fellowship. Lasswell was influenced by John Dewey, George Herbert Mead, and Robert Park during his studies at Chicago.

Eventually, Lasswell's eclecticism led to a crisis in his career at the University of Chicago, where he had been a Ph.D. student (1923–1926) and then a very productive faculty member for fifteen years. Chicago's president, Robert Maynard Hutchins, was prejudiced generally against the social sciences and especially against empirical social scientists (this attitude of Hutchins's contributed to the downfall of the Chicago School of Sociology after 1935). Lasswell was not only an empiricist, but he dabbled in Freudian theory, utilized content analysis (some regard him as the founder of content analysis, which Lasswell called newspaper "space analysis") to investigate propaganda effects on public opinion, and resolutely refused to be pigeonholed (Schramm, 1985, p. 203). President Hutchins vetoed Lasswell's promotion to full professor in 1938, and so the rejected scholar packed his books and research notes on a moving van and headed east. While en route the truck and all of its contents burned. After a dozen years at Chicago, where he had trained some of the ablest political scientists of our day (such as the late Ithiel de Sola Pool and Nobel laureate Herbert Simon), Lasswell was not to train another in the remaining forty years of his life, as he taught at the Yale Law School, which did not award Ph.D. degrees (Schramm, 1985, p. 203).

Lasswell remained an active scholar, however, writing over 6 million words of scholarly publication during his lifetime. His three-volume *Propaganda and Communication in World History* was in press at the time of his death in 1980. The Stanford University Library contains fifty-seven different books that Lasswell authored, co-authored, or edited! Out of this Guinness's record production of scholarly writing, most modern communication students know Harold Lasswell for only one sentence, written in 1940, in which he defined the scope of communication research: "*Who* says *what* in *which channel* to *whom* with *what effects?*" (Lasswell, 1954, pp. 37–51).[14] Clearly, these five interrogatives headed communication research in the direction of studying effects. Content analysis was the appropriate tool used to investigate message content (the "*what*"), media analysis explored the issue of "*which channel*," and audience surveys provided understanding of the "*to whom.*" Mass communication in the decades since Lasswell's sentence has largely ignored the "*who*," that is the investigation of which individuals and institutions own and control the mass media (Czitrom, 1982, p. 132).

KURT LEWIN: GATEKEEPING

Two of Hitler's direct contributions to American communication science were Kurt Lewin and Paul F. Lazarsfeld. Lewin was a German Jew who had studied psychology at the University of Berlin, and became a professor of social psychology there. He fled Naziism in 1933, landing in an unlikely spot, the University of Iowa, where he became a remarkable professor, attracting some of the brightest doctoral students of that generation. He was a very enthusiastic teacher, a charming personality, and soon had a cultlike following. Lewin lectured in a kind of fractured English; Wilbur Schramm, who was at Iowa at the same time as Lewin, remembers him pacing in front of the chalkboard, face flushed with excitement, challenging his doctoral class: "Vot haf ve fergassen?" (Schramm, 1985, p. 206). What variables had been forgotten (in the typology under discussion)?

Kurt Lewin had a fabulous ability to bring out the best performance in other people, especially in intellectual matters. He led a weekly discussion session, which he called the *guasselstrippe* (his Iowa students translated this German term as "The Hot-Air Club"), at which anyone could present a theory or a research plan for consideration and debate. Margaret Mead, the famous anthropologist, who collaborated with Lewin in his World War II food experiments, said of him: "Kurt was like the fire around which other people gathered for warmth and light by which to read their own thoughts more clearly" (quoted by Marrow, 1969, p. 91). Lewin had a rare ability to sense which research problems were worthy of investigation, and which were nonsense. This judgmental ability was one reason why Lewin is considered the founder of modern experimental social psychology (Festinger, 1980, p. 2).

Lewin launched the field of group dynamics, focusing on problems of group communication as a means of understanding how individuals are influenced by the groups to which they belong. How do autocratic versus democratic leadership types affect group productivity? How does an individual's attachment to a group affect that individual's conforming to group norms? These and other social psychological questions were investigated by Kurt Lewin and his Iowa students, many of whom went on to dominate the field of social psychology: Alex Bavelas, Dorwin Cartwright, Leon Festinger, Ronald Lippitt.

Many of Lewin's research interests had an applied significance; in fact, Lewin's research funds mainly came from the Iowa Child Welfare Research Station at the university. The origin of his resources made them no less attractive to Lewin. He is known for his sensible dictum: "There is nothing as practical as a good theory." A book about Lewin's life and work by his protégé (Marrow, 1969) is appropriately titled *The Practical Theorist*. To Lewin, theory and research could be one; even applied research projects should test theories. There have been few sources of funding for basic research in communication science over the years, but relatively plentiful funding for the study of applied communication problems. So Lewin's dictum has proven a useful strategy to many who followed in his way.

Lewin's insistence that social science would win wider acceptance only if it contributed to social change (Coser, 1984, p. 25) fit well with the liberal, New Deal era of America in the 1930s, and during the wartime 1940s. But Lewin's emphasis on praxis put off the top officers of the American Psychological Association (APA), who felt that applying psychology to real-world problems would lower the academic prestige of the field. So Lewin helped found a professional association to study social issues. Unlike most other psychologists, he had one foot in the world of practice, and one in the world of theory. Further, Lewin originated the famous "T-group" summer training course at Bethel, Maine, which influenced the many later sensitivity training courses. Originally, the T-group process was heavily studied by Lewin and his colleagues, but eventually the research purposes got lost.

Contemporary communication scholars know Kurt Lewin best for his concept of *gatekeeping*, the process of controlling the flow of messages in a communication channel. During World War II, the U.S. government promoted the public consumption of sweetbreads (undesirable cuts of meat, such as the heart, tripe, and intestines); Lewin and his students conducted a series of small group experiments with Iowa City residents in which different appeals were made to encourage eating sweetbreads. The Iowa social psychologists found that housewives acted as gatekeepers for the unpopular foods; unless a housewife decided to promote sweetbreads to her family, it was unlikely that her husband or children would eat them. In Lewin's (1947) last article before his untimely death, he theorized about the gatekeeping process in a communication system. Lewin's students (for example, David Manning White [1950]) and other communication scholars soon were studying gatekeepers in mass media institutions, such as the news wire editors of newspapers, who control the flow of national and international news into a local paper. Today, the concept of gatekeeping is widely used in communication science, particularly in organizational communication.

Shortly before his death from a heart attack in 1947, Lewin and several of his Iowa students moved to MIT, where he launched a Research Center for Group Dynamics. Lewin raised the funds for this institute from foundations and private sources. With Lewin's death, his Center moved to the University of Michigan, where it continues at present. Small group communication studies are today relatively passé, but Lewin's theoretical contributions to communication science live on.

CARL HOVLAND: PERSUASION RESEARCH

The only one of the four founders of communication research without a strong European connection was Carl Hovland. He was neither a refugee scholar like Lewin and Lazarsfeld, nor had he studied in Europe as had Lasswell. Hovland took his Ph.D. in experimental psychology at Yale University, where he was a protégé of Clark Hull, the great learning psychologist, and began his career there, running rats in learning experiments. In fact, Hovland was the "boy wonder" of experimental psychology; at age thirty-one he had had more articles published in the prestigious *Journal of Experimental Psychology* than any other psychologist of *any* age. But like the other founders of communication science, Hovland's career took an unexpected turn. World War II began and the young Professor Hovland was called to Washington to work for the U.S. Department of War (now Defense). He was assigned to study the effects of troop orientation films on fighting morale, a seemingly humdrum piece of highly applied work.

Not, however, in the hands of Carl Hovland. He designed experiments on the army training films so as to test theories about source credibility, one-sided versus two-sided presentation, the use of fear appeals, and immediate versus delayed effects. The dependent variable in these studies was persuasion (that is, the degree of his soldier-respondents' attitude change). The transition from rats to humans as experimental subjects converted Hovland from an experimental psychologist to a social psychologist with a fundamental interest in communication effects.

When the war was over, Hovland returned to Yale, with his data, to found the Yale Program in Communication and Attitude Change. He summarized his wartime studies in *Experiments on Mass Communication* (Hovland and others, 1949). Soon a stream of books and monographs, such as *Communication and Persuasion* (Hovland and others, 1953), flowed out of his research program. Hovland's studies infected communication research with a focus on effects (measured by the degree of attitude change). Persuasion research was to become a very popular type of communication research, and it is to this day. One of Hovland's former Ph.D. students, William McGuire, the contemporary intellectual leader in persuasion research, estimates that 1,000 new publications on persuasion research appear each year (McGuire, 1981, p. 43).

PAUL F. LAZARSFELD: TOOLMAKER

Of the four founders, Lazarsfeld had the greatest influence on the direction taken by communication research, and so his career and his methodological/theoretical contributions deserve the greatest detail here. Lazarsfeld had earned his Ph.D. in mathematics at the University of Vienna in the mid-1920s. Then, for a decade, he taught on the faculty, and directed a social science research institute, at the University of Vienna. This institute represented a common meeting ground between the academy and government and industry, who sponsored the research projects carried out by Lazarsfeld and his staff. Although the Vienna institute did not conduct communication research (in fact, Lazarsfeld was not yet interested in the topic), it later served as a model for communication research centers on the other side of the Atlantic.

Paul Lazarsfeld, a Jew, left for America in 1933, as the Nazi shadow of anti-Semitism began to fall across Austria. As related previously, Lazarsfeld first became director of the Office of Radio Research at Princeton, whose work was sponsored by the Rockefeller Foundation. In 1939, he moved this institute to Columbia

University in New York and renamed it the Bureau of Applied Social Research; Lazarsfeld was its director and was appointed professor of sociology. At this point, he saw himself mainly as a research methodologist, as a toolmaker for the social sciences. He matched his interests in technique with Robert K. Merton, an eminent sociological theorist at Columbia; their collaborative researches on a variety of issues were to become legend over the next several decades.

By the late 1930s, radio was having an important influence on American life, and this explains why the Rockefeller Foundation sponsored the Princeton and Columbia research programs. Lazarsfeld was originally attracted to radio research in 1937 because "Radio is a topic around which actually any kind of research method can be tried out" (Lazarsfeld, quoted in Czitrom, 1982, p. 129). Soon, however, the Viennese scholar began to see his radio research as a direct step toward a more general kind of communication research. At first, Lazarsfeld and his research staff primarily conducted audience research intended to simply determine the size and characteristics of the radio-listening audience. Then the radio studies began to turn toward the effects of radio.[15] These radio researches were sponsored by the Rockefeller Foundation with CBS also providing considerable funding to Lazarsfeld, who maintained a close friendship with Frank Stanton, the CBS director of research and, later, CBS president. The intimate relationships of the U.S. media industry with Lazarsfeld's bureau led to charges of an Establishment bias in his communication research by Daniel Czitrom (1982, pp. 127–139) and Willard Rowland, Jr. (1983, pp. 55–92). Under Paul Lazarsfeld's entrepreneurial hand, communication research moved from being an individual scholarly activity to becoming the enterprise of a large research team. This change inevitably moved communication research closer to the Establishment, on whom it depended for funding.

Lazarsfeld soon turned to investigating media other than radio, and to studying special local community contexts in which these media had effects. With sponsorship from the Time-Life Corporation, Lazarsfeld mounted an important investigation of the 1940 presidential election. Bureau of Applied Research investigators conducted repeated interviews with a panel of 600 citizens in Erie County, Ohio, in order to determine the role of the mass media in changing voting choices. Little direct effect was found, leading Lazarsfeld and others (1944) to posit a "two-step flow of communication" in which the media influenced opinion leaders, who then influenced other individuals. The Erie County project launched a research tradition on the "limited effects" of the media, a view that considered the mass media as just one among various other influences on human behavior, and not a very important influence at that (Chaffee and Hochheimer, 1985).

In the mid-1940s, MacFadden Publications sponsored a Bureau project on the role of the mass media and interpersonal influence on the consumer decisions of 800 women in Decatur, Illinois, later reported in the book *Personal Influence* (Katz and Lazarsfeld, 1955). As the book's title implies, the media were found to be much less influential than were interpersonal networks. The ultimate statement of the limited effects viewpoint was Joseph Klapper's *The Effects of Mass Communication* (1960), a synthesis of media effects studies that was originally completed in 1949 as Klapper's doctoral dissertation, carried out under Lazarsfeld's guidance and sponsored by CBS (Rowland, 1983, p. 72).[16]

Like most of the other early ancestors of communication science, Lazarsfeld was strongly committed to social change; he felt society was an imperfect system, in need of a major tune-up. Lazarsfeld had been a Socialist radical in his Vienna days; in fact, he started his research institute primarily to determine why socialism was so unsuccessful in Austria (Coser, 1984, p. 112). In America, Lazarsfeld considered himself a "Marxist on leave," a strongly reform-minded social scientist who thought that increased methodological sophistication was the direct route to tuning up society (Coser, 1984, p. 119). Paradoxically, the émigré scholar from Socialist Vienna became a major force in institutionalizing social research in capitalist America (Pollak, 1980). In contemporary times, Lazarsfeld has become a favorite target for attack by neo-Marxist scholars of the critical school of communication research.

More than any other scholar, Paul Lazarsfeld carried communication research in the direction of empiricism. Increasingly sophisticated research methods of sampling and measurement earned communication science the gradual respect of other social sciences (communication scholars often simply "out-empiricized" them), and gave communication research a certain degree of power with policymakers. But the heavy empiricism came at a cost, as several critics were to point out (Czitrom, 1982, p. 140). One of these was Robert S. Lynd, a Columbia University sociologist, who warned of the seductive quality of empiricism in the social sciences:

"To carry it on one usually places oneself inside the going system, accepts temporarily its values and goals, and sets to work at gathering data. . . . He tends to be drawn deeper within the net of assumptions by which the institution he is studying professes to operate" (Lynd, 1939, p. 120). Empiricism can keep one from asking such troublesome questions as how adequate the system of study really is, and what might be a better alternative.

The other critic was the leftist sociologist C. Wright Mills (1959), also a colleague of Lazarsfeld's at Columbia, who bitingly disapproved of his research style, which Mills called "abstracted empiricism." The range of social science, Mills argued, was unduly limited by confining the word *empirical* to mean statistical information about present-day individuals (Czitrom, 1982, p. 142).

The four founding fathers cross-fertilized the new communication science with the relevant social sciences disciplines, provided it with a sophisticated, if narrow, methodology, and grounded it in a heavy empiricism. Most of all, they steered the field into a solitary focus on communication effects. From the hindsight of today, these shortcomings of the founders are evident. But so are their crucial contributions in having made the newest field of social science become a reality.

Wilbur Schramm: Institutionalizer

The four founders were in a key location in the network of social scientists of their day: They taught at elite universities, published from the center of their disciplines, and thus legitimized a communication perspective on the range of human behaviors (voting, consumer behavior, attitude change, public opinion formation, group processes, etc.) that they studied. Without Lasswell, Lewin, Hovland, and Lazarsfeld, communication science could never have achieved its present status.

Nor could it have caught on without one other individual: Wilbur Schramm. It was his vision, "more than anyone else, that communication could become a field of study in its own right" (Paisley, 1985, p. 2). Schramm launched the first institutes of communication research (at Illinois in 1947 and at Stanford in 1956). Others soon followed. These institutes conducted scholarly research, trained the new Ph.D.s in communication science, and helped found communication research journals and scientific associations. In short, the institutionalization of communication science allowed an integration of theories and methods for attacking communication research problems. Gradually, over the next thirty years, communication science began to emerge as a discipline (Rogers and Chaffee, 1983). So Wilbur Schramm was the institutionalizer of communication research. In the 1940s and 1950s, this was a crucial role for the take-off of communication science.

Schramm was born and attended college in the small town of Marietta, Ohio; studied for his master's at Harvard; and earned his Ph.D. in American literature at the University of Iowa, where he then joined the faculty. Schramm taught creative writing at Iowa in the 1930s, where he directed the Iowa Writers' Workshop at the University of Iowa; he was himself a highly-successful fiction writer. Schramm was awarded the 1942 O.Henry prize for fiction for his short story, "Windwagon Smith," about a farmer with a flying tractor. His career direction changed radically, however, prior to World War II when he undertook a two-year post-doctoral course in psychology and sociology. During the war, Schramm served in Washington, D.C., in the Office of War Information, where he came in close contact with Lasswell, Hovland, and other social scientists of communication (Rowland, 1983, p. 82). In 1943, Schramm returned to Iowa City as director of the School of Journalism.

Four years later, he was lured to the University of Illinois to found the Institute of Communications Research, the first of its kind. In 1950, the first Ph.D. candidates in mass communication were admitted at Illinois, and Schramm became dean of the Division of Communication (which included journalism, speech communication, and other units of applied communication). During his Illinois years, Schramm published several textbooks on communication, notably *Mass Communications* (1949) and *The Process and Effects of Communication* (1954). He also arranged for the influential Claude Shannon/Warren Weaver book, *The Mathematical Theory of Communication* (1949), to be published at Illinois.

In 1956, Schramm moved to Stanford University to found the Institute for Communication Research, which, like the Illinois institute, was patterned after the Lazarsfeld model (Rowland, 1983, p. 83). Both institutes conducted communication research—often supported by grants and contracts with the federal government and private foundations and industry—and gave Ph.D. training. Rowland concludes his review of the rise of communication science by stating: "It is difficult to underestimate Schramm's role in the postwar development of communication research" (Rowland, 1983, p. 84). Perhaps Schramm should be considered a fifth founder of communication science. His books in the 1950s and 1960s laid out new fields for communication study: effect of television on children, communication's role in the development of Third World nations, and the nature of international communication flows among countries. Certainly, Schramm's theoretical contributions to communication science were no less than those of the other four founders.

Until Schramm, scholars came to study communication, tarried for a few years, and then moved back to their own parent disciplines. The four founding fathers were classic illustrations of this in-and-out venturing. But Wilbur Schramm came to communication science, and he stayed.

No other scholar did more to make the vision of communication science as a discipline become reality in the sense of its intellectual and institutional basis in American universities. During the past several decades, hundreds of university departments of communication have been launched in the United States. Some arose out of existing departments of speech, journalism, and other university units emphasizing a professional and/or humanistic kind of communication study. Other communication departments in American universities were created anew. All center around a behavioral science approach to understanding human communication, they employ as professors Ph.D.s trained in this intellectual viewpoint (thus gaining academic respectability and an ability to attract good students).

Were any of the four founding fathers of communication research alive today, they would be very surprised at the great numbers of university students now attracted to study communication. At most American universities, communication majors outnumber those in the other social sciences. The communication major, both for undergraduate and graduate students, has come in vogue in America, especially in the 1970s and 1980s. Likewise, the volume of mass communication research is impressive; for example, about 115 Ph.D. dissertations in mass communication are completed each year. This is about 1/250 of all dissertations in the U.S. (Paisley, 1985).

As communication science in the United States moved from being a "field" toward becoming a "discipline," during the era of intellectual consolidation and integration since about 1970, some of the intellectual diversity in early communication research has been lost. Communication research's strong academic connections to the other social sciences have weakened as the new field took on independent status. Part of the old excitement of the early days of communication as a field inevitably disappeared. It is now much easier for communication scholars to talk with each other and be understood in light of the increased homogeneity of their concerns, their common backgrounds in a shared training experience, and the rise of scientific journals reporting their research. The gradual accumulation of theoretically oriented, empirically based research findings is thus facilitated. In this way, the communication discipline progresses.

Wilbur Schramm should be pleased with the accomplishments of the discipline he helped launch.

Communication Technology and Communication Science

The history of communication research shows that its intellectual concerns were highly related to each of the new communication technologies that came, in turn, on the American scene. In the early part of the present century, newspapers were the dominant mass medium in the United States. Scholars such as John Dewey and Robert Park were involved in theorizing about, and in conducting research on, newspapers. When film began to command large national audiences in the 1920s and 1930s, coordinated programs of research (for example, the Payne Fund studies) were mounted to determine the possibly harmful effects of film-viewing on children and youth. Radio was next on the scene (in the 1930s), and Paul F. Lazarsfeld conducted communication research on this new technology. In the 1950s, it was television, with Wilbur Schramm pioneering in studies of TV's effects on children. Today, an increasing number of contemporary communication scientists conduct research on the social impact of such interactive technologies as computers.

Certainly one of the reasons for the support of communication science over the decades is its important contribution in understanding the *effects* of new technologies. No one could claim that the study of communication effects is worthless; effects questions are simply fundamental to the nature of human communication. On the other hand, few observers could argue that the very heavy concentration of U.S. communication scholars on effects is not overdone. Further, too much of the effects research has followed an oversimplified, one-way model of communication that ignored the context of communication, and falsely distorted its phenomenological nature by overlooking the inherent subjectivity of human interaction. Clearly we do not need more of the same kind of past communication research in the future. Effects issues *can* be studied in ways that do not ignore context or subjectivity. And of course, there are many important communication research issues to be studied other than communication effects.

The immediate imperative for shifting the direction of communication research comes from the interactive nature of computer-based communication systems of the 1980s. Linear models cannot be followed by scholars when the media are no longer one-way. Simple effects cannot be studied effectively when the impacts of the new media are complexly variable for each individual participant in a communication system. False dichotomies such as mass media versus interpersonal face-to-face communication channels, which have for too long divided the discipline, will fade in the face of the new machine-assisted interpersonal communication.

Many communication scholars have been strong technological determinists, assuming that new communication technologies are one of the important causes of social changes in society. Initially, scholars have greeted a new communication technology with overblown optimism about its potential effects. Later, communication researchers often have directed their attention to the negative effects of a new technology. Frequently, priority attention has been given to children and youth as an audience for such negative effects because these younger people usually are more accepting of each new technology, and thus are a priority audience for assessing the social impacts of a new technology. Often the impacts of a communication technology are greater on youth than on older audiences. Children have often served as a kind of litmus paper for the effects of new communication technologies, as we see in the case of both television and computers.

It is probably no accident that communication science really took off with the electronics era, as marked by use of the first computer in 1947. But computers and communication have only come together very recently, in the 1980s. The full intellectual implications of this merger are not yet widely grasped by communication scholars, nor can they be fully understood for some years.

In further chapters of this book, we hope to detail just what the new communication technologies will mean for communication science, theoretically and methodologically.

Perhaps it is appropriate that I close this chapter on a personal note, as it began. My personal research interest in new communication technology began only very gradually and grew slowly in the face of an antitechnology attitude developed while I was a doctoral student in the 1950s. Then I was taught that communication technology should be the concern of electrical engineers and physicists, but that it had little significance for social scientists like me.

This basic belief in the irrelevance of communication technology continued through the 1960s, when, like most other Americans, I read the popular writings of Marshall McLuhan, the Canadian scholar of communication technology, who became well-known for his dictum that "The Medium is the Message." In such books as *Understanding Media*, McLuhan (1965) claimed that such communication technologies as television were "extensions of man," broadening one's senses so as to enable one to reach out for information into the environment. Through the 1960s, I continued to think that communication technology was a non-factor in communication research. At least it was a non-factor in *my* communication research.

My resistance began to crack in the early 1970s. I remember expressing my surprise to Ed Parker, then a professor of communication at Stanford University, that he had become involved in research on computers and communication satellites. Parker, who was to co-found a successful Silicon Valley satellite communication company, of which he is presently vice-president, remarked that technology variables were where the action was in communication research. He pointed out to me that in my research—a series of field experiments in Third World nations on development communication—I was fortunate to explain 10 or 20 percent of the variance in my dependent variables (of knowledge about, and adoption of, health and agricultural innovations). In contrast, Parker explained 70 to 80 percent of the variance in such behavior variables in his study of two-way satellite radio communication linking Eskimo and Indian villagers in Alaska with medical expertise. Instead of accepting communication technology as a given, Parker considered it as a variable that communication scholars could control and influence. But I still had doubts that the study of communication technology was a legitimate concern of communication scientists.

A couple of my colleagues, however, were convinced. One was Fred Williams, the founding dean of the Annenberg School of Communications at the University of Southern California in 1973. Williams wisely positioned his School's research and teaching to concentrate on the new communication technologies. Within a decade, many other academic leaders in U.S. communication science had swung around to the intellectual position pioneered by Ed Parker and Fred Williams.

Today, almost every school or department of communication at an American university employs at least one faculty member specializing in new communication technology. Research on this topic is a high priority for communication scholars. Communication majors now are required to enroll in computer programming courses and to utilize computers in news writing and in television and film production, as well as in data-analysis.

During the 1980s, investigation of the diffusion and social impacts of new communication technologies has become my research interest.

Notes

1. Pragmatism was founded by Charles Sanders Peirce and carried forward by the Harvard psychologist William James, as well as by Dewey.

2. There is a considerable literature about the Chicago School of Sociology: Robert Faris (1970), James Short (1971), Paul Baker (1973), James Carey (1975), Fred Matthews (1977), P. Jean Frazier and Cecile Gaziano (1979), J. David Lewis and Richard Smith (1980), Daniel Czitrom (1982), Martin Bulmer (1984), Norman Denzin (1984), and Henrika Kuklick (1984).

3. Is communication science a discipline today? Certainly it is by the usual criteria of the number of scholars and of university departments that are involved, their intellectual integration by means of conferences and scientific journals, and their self-identification. On the other hand, a wide diversity in theoretical viewpoint within communication science leads some scholars to call it a "multidiscipline" (Paisley, 1985).

4. A *paradigm* is a scientific approach to some phenomena that provides model problems and solutions to a community of scholars (Kuhn, 1970).

5. A further role of Wilbur Schramm in promoting Shannon's theory was his article on the implications of information theory for mass communication research (Schramm, 1955).

6. Personal communication with Harriet P. Stockanes of the University of Illinois Press, March 15, 1985.

7. Today, communication research is concerned with determining whether machine-assisted interpersonal communication (such as that occurring via computer teleconferencing in computer networks) is different from face-to-face interpersonal communication. The findings to date indicate that machine-assisted communication is less effective in conveying socio-emotional content, although it can successfully exchange technical content (Chapter 2).

8. Thus Shannon's communication model was not very closely in line with the Latin root *communico*, which means "share."

9. Schramm concluded his article about Shannon's information theory and communication research by stating: "Of all the potential contributions of information theory to mass communication, perhaps the most promising is the study of communication networks" (Schramm, 1955). Unfortunately, research on networks has only began to command the attention of communication scientists in recent years.

10. Norbert Wieners drew on various academic ancestors in stating his cybernetic theory: "His role in cybernetics was not only that of an innovator but also that of a publicist, synthesizer, unifier, popularizer, prophet, and philosopher" (Heims, 1980).

11. Wiener's student, Karl W. Deutsch, was not so influential in communiction science as Schramm. Deutsch (1963) carried his teacher's cybernetic thinking forward into a focus on political communication networks in what Deutsch called "the nerves of government."

12. About the same time that Lasswell was analyzing propaganda and public opinion, the influential journalist (and friend of twelve different U.S. presidents) Walter Lippmann was writing such books as *Public Opinion* (1922). Lippmann claimed that public opinion was created when the mass media conveyed emotionally loaded stereotypes of news events. These helped form pseudo-environments or "pictures in our heads," which came to represent the perceived reality of news events that had occurred at a distant time and place. Lippmann was a practicing journalist, not a university professor, and although his ideas had some influence on the emerging field of communication science, he does not rate as one of the founders of the field.

13. The Critical School traces to its Marxist beginnings as the Institute

for Social Research in Frankfurt, Germany, in the early 1930s. Today, critical scholars in Europe and America set themselves off from
empirical scholars by objecting to the effects-oriented, empirically
minded nature of most communication research (Rogers, 1985a, pp.
219–235; Coser, 1984, pp. 85–101).

14. This statement first appeared in a 1940 memo entitled "Research in
Mass Communication," signed by ten scholars including Harold
Lasswell and Paul Lazarsfeld (Czitrom, 1982, p. 217). The memo, intended only for private circulation, summarized the conclusions and
recommendations of a meeting sponsored by the Rockefeller Foundation that explored the future of communication science. The five-
question definition of communication research is credited to Lasswell
because he first stated it at the meeting (Rowland, 1983, p. 64) and
was first to publish it (Lasswell, 1954, pp. 37–51).

15. A series of books reported the radio research by Lazarsfeld and his colleagues, with the best summary provided by Paul F. Lazarsfeld and
Frank Stanton (eds.), *Communication Research, 1948–1949* (1949).

16. The limited effects model was often described at the time as representing a rejection of the "hypodermic needle model," the notion that the
mass media had direct, immediate, and powerful effects on an audience. Recently, however, several communication scholars have
questioned whether such a hypodermic needle model had ever been
actually stated (Chaffee and Hochheimer, 1985). Perhaps the
hypodermic needle model was just a strawman, created as a convenient device for Lazarsfeld and his colleagues to destroy. In any event,
Lazarsfeld generally found that the mass media's effects were much
less than had been expected. It is a tribute to Lazarsfeld's entrepreneurial salesmanship that he could continue to secure the financial support of media institutions for research that found limited effects.

CHAPTER 4

Adoption and Implementation of Communication Technologies

"Learning to use a computer is much more like taking up a musical instrument than following instructions on how to use an electrical applicance, such as a toaster."

Charles Rubin, August 1983

One of the two main areas of research on new communication technologies deals with how these innovations are adopted and implemented by users. The other main topic of research, the impacts of the technologies, is moot until adoption has occurred. Research on the adoption of the new media is based on a contemporary application of a well-researched theory with a long history, that of the diffusion of innovations. However, there are several special aspects of the application of this theory to the case of the new communication technologies. For instance, their interactive nature means that the value of the innovation to the adopter becomes greater with each succeeding adoption (as an extreme case, an electronic messaging system is worthless to the initial adopter until there is at least a second adopter). Further, the degree of *use* of a new communication technology becomes an important variable, in addition to whether or not adoption has occurred.

Here are some main questions that will be discussed in the present chapter.

- Who adopts the new communication technologies? What are the characteristics of the adopters?
- How rapid is their rate of adoption?
- How does adoption of one new medium affect the adoption of others? For example, do people with home computers also have VCRs? Is there a key technology that, once adopted, triggers adoption of other communication technologies?
- What is the innovation process, including implementation, of a new communication technology in an organization?

Diffusion of Innovations

For the past forty-five years and through 3,500 research publications, a model of the diffusion of innovations has guided investigations of the spread of new ideas. Diffusion research is conducted by scholars in such fields as anthropology, communication, education, geography, marketing, rural sociology, sociology, and several others. Although this multidisciplinary research is carried out on a variety of innovations (such as new consumer products, agricultural techniques, and new medical drugs) as they spread among diverse audiences, diffusion studies are all guided by a common theory. The wide applicability of diffusion theory explains why so many diffusion researches continue to be conducted (McAnany, 1984). Here we summarize the main elements of this framework, and discuss its applications to new communication technologies.

The main elements in the diffusion of new ideas are: (1) an innovation, (2) that is communicated through certain channels, (3) over time, (4) among the members of a social system (Figure 4–1). An *innovation* is an idea, practice, or object perceived as new by an individual or other unit of adoption. The characteristics of an innovation, as perceived by the members of a social system, determine its rate of adoption. Five attributes of innovations are: (1) relative advantage, (2) compatibility, (3) complexity, (4) reliability, and (5) observability.

A *communication channel* is the means by which messages get from one individual to another. Mass media channels are more effective in creating knowledge of innovations, whereas interpersonal channels are more effective in forming, and in changing, attitudes toward a new idea, and thus in directly influencing the decision to

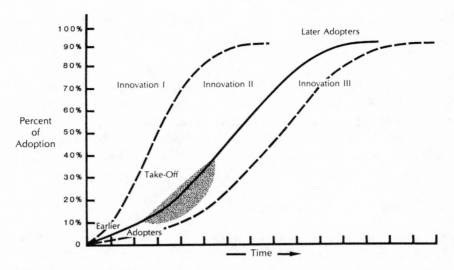

Figure 4-1. Diffusion Is the Process by Which (1) an *Innovation* (2) Is *Communicated* Through Certain *Channels* (3) Over *Time* (4) Among the Members of a *Social System*

Here we see that innovation I has a more rapid rate of adoption than innovation III. Each of the three diffusion curves are S-shaped, with a take-off in the rate of adoption at about 10 to 25 percent adoption. This take-off occurs when the opinion leaders in a system adopt the innovation, thus activating interpersonal diffusion networks.

SOURCE: Rogers (1983, p. 11). Used by permission.

adopt or reject a new idea. Most individuals evaluate an innovation that they are considering adopting, not on the basis of scientific research by experts, but through the subjective evaluations of near-peers who have previously adopted the innovation. These near-peers thus serve as social models, whose innovation behavior tends to be imitated by others in their system.

Time is involved in diffusion in (1) the innovation-decision process, (2) *innovativeness*, the degree to which an individual or other unit of adoption is relatively earlier in adopting new ideas than other members of a social system, and (3) an innovation's rate of adoption. The *innovation-decision process* is the mental process through which an individual or other decision-making unit passes from first knowledge of an innovation, to forming an attitude toward the innovation, to a decision to adopt or reject, to implementation of the new idea, and to confirmation of this decision (Figure 4-2). Five

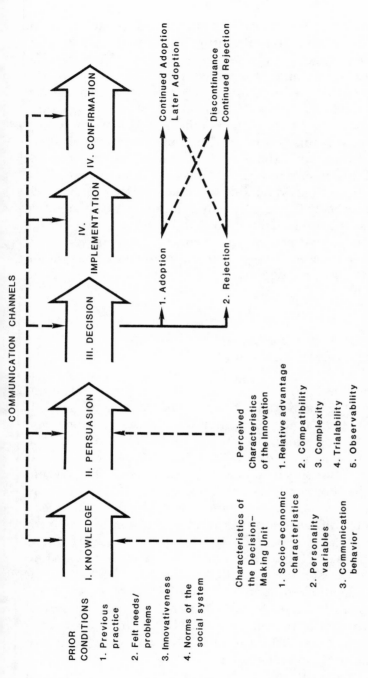

Figure 4–2. A Model of Stages in the Innovation-Decision Process

The *innovation-decision process* is the mental process through which an individual or other decision-making unit passes from first knowledge of an innovation, to forming an attitude toward the innovation, to a decision to adopt or reject, to implementation of the new idea, and to confirmation of this decision.

SOURCE: Rogers (1983, p. 165). Used by permission.

119

steps in this process are: (1) knowledge, (2) persuasion, (3) decision, (4) implementation, and (5) confirmation. An individual seeks information at various stages in the innovation-decision process in order to decrease uncertainty about the innovation.

A *social system* is a set of interrelated units that are engaged in joint problem-solving to accomplish a goal. A system has *structure*, defined as the patterned arrangement of the units in a system, which gives stability and regularity to individual behavior in the system. The social and communication structure of a system facilitates or impedes the diffusion of innovation in the system. For example, an *opinion leader* is an individual able to influence other individuals' attitudes or overt behavior informally in a desired way with relative frequency. The opinion leaders embody the norms of a system, the established behavior patterns for the members of a social system. If the norms of a system are favorable to an innovation, the opinion leaders are more likely to adopt it, and other individuals will tend to follow their lead.

What Is Special About the Diffusion of Communication Technologies?

Diffusion theory has been built up gradually through investigations of a wide variety of innovations: the snowmobile among the Skolt Lapps; gammanym, a new drug, among medical doctors; and modern math among American schoolteachers, for example. The process of diffusion across these different innovations shows a great deal of regularity. But only in quite recent years have scholars begun to investigate the adoption of new communication technologies. Are there certain distinctive qualities of the new media that would lead us to expect their diffusion to differ from that generally found for other new ideas?

1. A *critical mass* of adopters of an interactive communication technology are necessary for the utility of the new idea to be sufficient for an individual to adopt. The usefulness of a new communication system increases for all adopters with each additional adopter. An extreme illustration is provided by the first individual to adopt a telephone in the 1870s; this interactive technology had zero utility until a second individual adopted.

The critical mass dimension of the interactive communication technologies is a crucial factor in their adoption and use. The critical mass aspect of the new media traces from their interactive quality. If these innovations were not themselves communication channels, of a special kind, the critical mass dimension would not be involved. Especially in the first stages of an innovation's diffusion, the critical mass dimension is a negative influence, slowing the rate of adoption. Past research shows that a computer-based messaging system quickly falls into disuse if there is a low likelihood of finding a desired individual on the system.

2. The new communication media are *tool technologies*, representing techniques that can be applied in a variety of ways to diverse situations. Such tool technologies as computer-based innovations are frequently characterized by a relatively high degree of re-invention. This concept came to be recognized only in diffusion research within the past decade; prior to that time it was assumed that an innovation was a standard quality, unchanging as it diffused. Now we realize that assumption is an oversimplification.

Re-invention is the degree to which an innovation is changed or modified by a user in the process of its adoption and implementation (Rogers, 1983, pp. 16–17). The frequent occurrence of re-invention tells us that the adoption of the new media is often a very active behavior in which the adopter customizes the innovation to fit his or her conditions, rather than just the passive acceptance of a standardized innovation. In a California study of the diffusion of home computers (described later in this chapter), the adopters typically went through a period of several weeks of frustrated problem-solving and information-seeking after their initial purchase. They gradually learned how to properly use their computer, and to apply this tool to various tasks, such as word-processing, accounting, playing video games, etc. Eventually, the respondents in the California home computer study became satisfied adopters, and enthusiastically recommended the innovation to their peers. Some of the adopters became almost addicted to using their computers.

One should not overlook the fact that the adoption of new communication technologies are seldom just a push-pull, click-click type of behavior. The individual must become considerably involved with the innovation. "Adoption" seems a much more appropriate term for this highly involving kind of behavior than "acceptance."

3. The dependent variable[1] in diffusion studies of the new media is often the degree of *use* of the innovation, rather than just the deci-

sion to adopt or even *implementation* (putting the innovation into use). Early diffusion researchers, like Ryan and Gross (1943) in their study of the diffusion of hybrid seed corn among Iowa farmers, simply asked their respondents whether or not they had decided to adopt the new idea, and, if so, when they had decided to adopt. *Adoption* was defined as the decision to make full use of an innovation as the best course of action available. For the Iowa farmers, adoption amounted to a decision to plant all of their corn acreage in hybrid seed. Until the 1970s, adoption served satisfactorily as the main dependent variable in diffusion studies, which generally were designed to determine the characteristics of individuals (1) who had adopted versus rejected an innovation, and (2) who had adopted relatively earlier than others.

Then, in the 1970s, diffusion scholars began to investigate the innovation process within an organization. This unit of adoption was more complex than the individual (for example, the Iowa farmer in the hybrid corn study). One or several individuals in an organization might make the decision to adopt an innovation, but then another set of individuals would be involved in implementing the new idea. Often a considerable period of time was required to put the innovation into use, and various problems would be encountered. Many times, an adoption decision was not fully implemented. Clearly the decision to adopt and the implementation of the innovation were not identical. So once diffusion studies began to investigate the adoption of innovations in organizations, the dependent variable often became implementation, instead of just the decision to adopt.

Implementation is frequently the dependent variable in studies of the diffusion of new communication technologies, rather than adoption. An illustration is provided by our case study of the implementation of microcomputers in high schools (later in this chapter). The degree of use of the new media frequently is the bottom line, as is illustrated in our case study of the uses of electronic mail in an organization (which is also found later in this chapter).

So there are several important ways in which the diffusion of communication technologies differ from the spread of other innovations: (1) the critical mass, (2) a relatively high degree of reinvention, and (3) the focus on implementation and use, rather than just on the decision to adopt. Further research on the diffusion of new communication technologies will serve to broaden the scope of diffusion theory.

Diffusion of Home Computers*

The microcomputer promises to become one of the most important innovations of recent decades in terms of the changes being caused in schools, places of business and homes. Perhaps the home computer will become a central artifact of the future Information Society, similar to the role filled by the automobile in the Industrial Society. One of the first investigations of the diffusion of microcomputers was conducted by the present author among a small sample of households in Northern California (Rogers and others, 1982). Since then, a dozen or more other investigations of home computer diffusion have been completed (Venkatesh and others, 1984; Dutton and others, 1986).

A *home computer* (or *personal computer*) is a microcomputer designed to be used by an individual, rather than an institution such as a business, educational organization, or government. A home computer is "personal" in that its use is intended for the owner (and perhaps his or her family). Home computers are microcomputers because a microprocessor is used as the central processing unit. The cost of a home computer ranges from a few hundred dollars to several thousand; the typical respondent in the California study spent $3,500 for his or her microcomputer and such peripheral equipment as a printer, disk drives, and, perhaps, a modem.

In our 1982 study, we conducted personal interviews with seventy-seven owners of home computers. These respondents were a nonrandom, but representative, sample of Silicon Valley home computer owners. They were technical and socioeconomic elites; 28 percent had Ph.D.s and 20 percent were scientists or engineers. Average personal income was $38,000, average age was thirty-six years, and 89 percent were males. Later diffusion studies showed that these characteristics typify the first adopters of microcomputers in the U.S. The strong relationship of formal education and of technical occupations with the adoption of home computers may be due to the active nature of the innovation, which involves the user to a high degree (quite unlike the adoption of TV in the 1950s, for example).

Sources/Channels of Communication About Home Computers

Interpersonal networks are more important than the mass media in creating awareness-knowledge of home computers. More than half of the respondents said they first heard about home computers from such interpersonal sources/channels as work associates, friends, and family members. Most previous diffusion studies report that the mass media are most important in creating awareness-knowledge of innovations, particularly at the first stages of the diffusion process (Rogers, 1983).

Microcomputers are one of the most widely advertised products in the U.S., yet despite the immense advertising campaigns mounted by IBM, Apple, and other microcomputer manufacturers, the diffusion of home computers is overwhelmingly a process of interpersonal networks. Evidence for this conclusion is provided by several of the findings:

- Home computer owners knew, on the average, about five other home computer owners prior to their purchase.
- Home computer owners talked to an average of nearly thirteen people per month about their home computers. Owners showed their computers to an average of about five persons per month.
- Home computer owners encouraged an average of eight other persons to buy a home computer.
- Computer owners received mainly positive messages from other computer owners before their purchase; positive messages outnumbered the negative by two to one.
- Despite the pro-innovation messages they received, two-thirds of the computer owners reported problems with their own computer. About a month or so was necessary after purchase before home computer owners felt comfortable using their computer. The respondents had expected to bring their new product home, plug it in, and begin using the computer to accomplish specific tasks. However, after their initial negative experience, most home computer users become very satisfied with their purchase. This initial frustration was a product of the high degree of re-invention that occurred for most new home computer owners as they fit the innovation to their particular situation.

Evidence of the need for information by many adopters of personal computers is the widespread participation in microcomputer users' groups. These local consumer associations are usually organized on the basis of the make of computer; for example, a Boston Apple users' group, a Des Moines IBM users' group, and others. Hundreds of thousands of microcomputer users have joined these groups, and continue to attend their monthly meetings. At these sessions, technical problems are discussed, software programs are copied and exchanged (perhaps illegally), and used computer equipment and peripherals are bought and sold. Mainly the users' groups flourish because they provide information and inforcement to new adopters of microcomputers. The existence of these users' groups suggests just how complex it is for most individuals to learn to use a microcomputer.

An overwhelmingly favorable type of interpersonal communication about home computers occurred through networks of near-peers. The satisfied owner is often a very convincing influence on neighbors and friends in the diffusion of home computers. Why are interpersonal net-

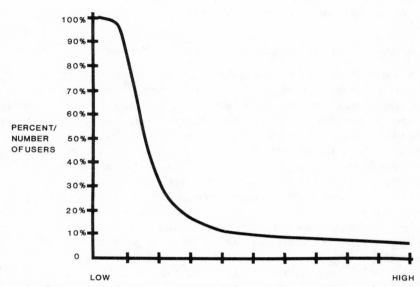

Figure 4–3. The Distribution of Degree of Use of a New Communication Technology Usually Consists of a Small Number of Heavy Users and a Large Number of Light Users

The general distribution of use shown here is abstracted from numerous studies of the patterns of use of new communication technologies. Whether the degree of use is measured in hours of use per day, week, or month, in the variety of uses, or in some other way, the distribution of use looks like the above. In the California study of home computer diffusion, many households used their microcomputer only six to fifteen hours per week. Average use for the entire sample was seventeen hours per week, but several respondents used their computer fifty to sixty hours per week (Rogers and others, 1982). A common finding in research on new communication systems is that only 10 percent of the users represent 50 percent of all uses, with the other 90 percent of users making up the other 50 percent of uses.

works so unusually important in the diffusion of home computers? Perhaps it is due to the high degree of enthusiasm that adopters feel for their computers. Many are "computer bores," insisting on discussing their microcomputers with friends, neighbors, work acquaintances, and even with strangers. Such missionary activity may be motivated by the involving nature of the home computer; it is a tool that an individual uses to accomplish a wide variety of tasks. Such utilization of a home computer demands that the individual becomes an active participant with the technology. This unique quality of microcomputers as a consumer product leads to the initial period of frustration as the owner learns to use the new equipment; later it creates a special kind of enthusiasm that leads the owner to activate interpersonal networks.

Consequences of Microcomputer Adoption in the Home

Consequences are the changes that occur to an individual or social system as a result of the adoption or rejection of an innovation. The typical home computer was used for seventeen hours per week. Home computer owners utilized their home computer about one-third of this time for playing video games, another third for word-processing, and all other uses constituted the remainder. Word-processing is a kind of automated typing that allows the user to compose and edit text as it appears on a computer screen. The text can be stored on a magnetic disk for future use, printed on paper, or transmitted to another computer over telephone lines. The main advantage of word-processing is that it makes writing and subsequent revisions more efficient because the computer performs many functions that the individual would have to perform manually on a typewriter.

The adoption of a home computer in a household leads to a decrease in the number of hours spent viewing television. In the average American household, the television set is turned on for seven hours per day. The California households of study decreased their television viewing by an average of thirty-four minutes per day, and one household reported an average decrease of four hours per day. This change represents a major shift in the use of time.[2] It suggests that home computers may be even more addictive than television.

Such computer addiction led to a variety of personal and family problems for our respondents. Computer users were reluctant to come to meals or go to bed ("Just a minute more until I finish what I'm doing here"). We heard complaints from "computer widows," whose husbands had become "computerholics." Parent-child conflict occurred over who got to use the computer. Home computer users also report getting less sleep because of the time they spend computing (Dutton and others, 1985).

After achieving a rather rapid rate of adoption for seven years, the diffusion of home computers slowed down in 1985, and the S-shaped diffusion curve leveled off with around 15 percent of American households adopting. The stock market prices of computer companies dropped, and these companies were forced to lay off thousands of employees. Several microcomputer companies went bankrupt.

Why did the diffusion curve for home computers level off? One basic reason is that there is not a functional need for a microcomputer in the home, except for a small minority of people (Caron and others, 1985). The word-processing function could be equally performed by an expensive typewriter, and the video games function could be filled more cheaply by a home video game machine. Other than word-processing and video game playing, the home computer is not really utilized very much. One of the main functions of home computers is to learn how to use a computer. The unique function of a home computer is to serve as a conduit in bringing information into the home, but only about 20 percent of home computer adopters had a modem in 1985 and utilized their equipment for computer communication. Why are home computer adopters so satisfied if their computer is not filling a real need? Perhaps because the individual who has adopted has to justify the considerable investment.

Additional uses of home computers will require much more software. Consumers also demand much lower prices for computers, perhaps half of what they were in 1985. Finally, because home computers are so complex to learn to use, simple and easier training techniques must be devised and made available. If these several prerequisites can be provided, the diffusion curve for home computers may take off again and begin to increase at a steady rate.

Innovation That Failed: The Context System at Stanford University*

In 1981, the computer center at Stanford University announced a new text-processing system called Context, which was promoted as a tool for scholarly writing. Context operated on one of the university's large computers, with each adopter of this system renting a computer terminal for about $300 per month. Context was a major improvement over such alternatives as a specialized word-processor (about a hundred or so were already in use at Stanford), which were relatively expensive to purchase, or text-processing on the Stanford mainframe computer, which entailed a more costly rental fee and where a writer's use was often interrupted for a period of time by huge data-analysis jobs.

Context was launched with considerable fanfare: stories in the weekly faculty newsletter, a videotape showing several prominent professors using the new service, and demonstrations. Scholars in the humanities, who had previously found little use for computers, were particularly expected to adopt Context. The first professors to begin using the new system were very enthusiastic, and it was hoped that the diffusion curve would take-off.

But only sixty-six of the 180 subscriptions Context needed to break even were achieved, and within two years of its upbeat launching, Context was discontinued. What went wrong?

Context was a victim of unfortunate timing. The designers of the new system had not anticipated that Stanford professors would purchase their own microcomputers, and prefer to use these stand-alone systems for word-processing. A microcomputer, disk drives, and printer could be purchased for the equivalent of one year's rental fees for Context. The computer center at Stanford, like those at most other large universities at the time, did not feel any responsibility for microcomputers, a decentralized technology that did not require centralized servicing by computer experts.

Further, Context proved to be very difficult to learn to use, especially by professors, such as the humanities faculty, who had not had previous computer experience. While Context was a superb word-processing tool in the hands of a skilled user, its sophistication unfortunately came at the cost of a high degree of complexity in its operations. Some potential users, especially older females, felt they could never learn to master Context. An investigation by Marcus (1985) showed that peer network communication among Stanford faculty and staff was crucial in diffusing Context. Most of these network links were with another individual in the same office or else very close by; only 6 percent of Context users knew an adopter in another department. The result was several pockets or clusters of adopters at Stanford, but not a campus-wide diffusion.

The Context system was canceled in late 1983, when Stanford University administrators thought the university was experiencing a budget crisis (a few months later, the perceived crisis mysteriously disappeared). Had they stuck with Context for another year or so, and absorbed the budget losses that it was generating, perhaps it might have eventually become widely diffused.

Smashing the ATM Wall[*3]

An *ATM* (automated teller machine) is an unmanned, automated electronic device that can dispense cash and handle such routine financial transactions as deposits or transfers between checking and savings accounts, through the insertion of a special banking card. ATMs represent a type of quasi-interactive communication in that the user interacts with a computerized bank teller programmed with a very limited range of responses.

During the late 1970s and early 1980s, about 70,000 ATMs were installed by U.S. banks, at a total cost of $1 billion. Originally, the ATMs were placed inside the bank; then they were moved outside to provide twenty-four-hour service, and finally ATMs were located in supermarkets, office buildings, and other convenient locations. One benefit of automated tellers is that they seldom make mistakes. The cost savings of the switch to ATMs is a tremendous boon to banks. Each machine costs $20,000 plus $10,000 for installation; but that is cheap compared to the $1 million cost of opening a new branch bank. The main advantage of ATMs for banks is the savings in labor. The Bank of America has 116 all-ATM banks, thus eliminating 1,400 teller jobs. Each transaction costs a bank 21 cents when it's done on an ATM, compared to 52 cents when the customer goes to a branch teller. So bankers are understandably enthusiastic about ATMs.

The only problem is getting bank customers to use them. By 1985, only one out of three U.S. bank customers used ATMs and this rate of adoption seemed to have stabilized. Accordingly, banks have tried various strategies to "smash the wall," the banker's expression for getting past the thirty-three percent adoption level. Positive incentives offered by banks to encourage adoption of ATMs include a small cash payment for using the new service, drawings for prizes, and lower monthly service fees to customers who use ATMs exclusively (a similar idea to paying a lower price for gasoline at a service station if you pump it yourself). A Pittsburgh bank even installed talking ATMs in order to make them seem more like human tellers.

When such carrots have not been effective in "smashing the wall" of ATM resistance, some banks have resorted to the stick. Generally, these negative incentives amount to making it more difficult or expensive for customers to use human tellers.

- A Boston bank put its tellers out of sight, and required customers to make an appointment in advance to see them; otherwise, they had to use ATMs.
- New York's Citibank in 1983 forced customers with less than $5,000 on deposit to use ATMs (this plan had to be quickly dropped in the face of customer reaction).
- Cleveland's Central National Bank required a customer who wanted to see a teller to insert his or her ATM card in a terminal, rather like taking a number at the meat counter of a grocery store.
- Numerous banks across the country use the simplest strategy of all: They severely understaff their banks with human tellers, so that the long lines of waiting customers force individuals to start using ATMs.

One reason for customer resistance to ATMs traces to their relative inaccessibility. For example, I have to drive ten minutes through Los Angeles traffic in order to reach my bank's nearest ATM; on this trip, I pass a dozen or more ATMs that belong to other banks, but I cannot use them. Much of the inaccessibility of ATMs is needless. If banks had joined together to develop a national network of compatible ATMs on the model of the telephone booth, this public service would have been much more convenient for bank customers to use. In Iowa, a state law required shared ATM terminals, and this system worked out well. But elsewhere, each bank insisted on having its own ATMs. And the customer suffers.

Perhaps the widespread resistance to using ATMs should tell us something about public perceptions of computers as a replacement for human service: that the socio-emotional function of interacting with our banker may be even more important to us than just the task function of making financial transactions.

Uses of an Electronic Mail System*

As mentioned previously in this chapter, investigations of the diffusion of the interactive communication technologies often seek to explain the degree of *use* of the innovation as their main dependent variable, rather than just the decision to adopt. A particularly fine example of this new genre of use studies is Charles Steinfield's (1983) research on an electronic mail system that had been operating for seven years in a large U.S. company that sold office products. The system had over 2,000 users, located in six U.S. cities and overseas.

The electronic mail system began in the company's R & D division, and then quickly spread throughout the organization (thus illustrating the critical mass dimension of diffusion). The typical user sent about two messages per day on the system, and received about the same number. An average of fifty minutes per workday was spent with the electronic mail system, a rather considerable period of time. As is usually the case, there were a few very heavy users of the system, and if these were removed, the average of two messages per day would drop to less than one message. The electronic mail system was free to employees for all possible uses, and this policy undoubtedly encouraged high use. No attempts were made by company management to discourage non-task-related communication, and in fact there was a great deal of such socio-emotional usage (especially by new employees).

The electronic mail system was integrated with word-processing and other computer functions at each work station. Employees retained most of their own computer files, but, as all work stations were linked in a network, large computer files could be stored in the company's mainframe computer. The central computer also routed electronic mail. Anyone with a "mailbox" on the system could send mail to any other individual. A message could be sent to a single individual, or to a *distribution list* (a group of individuals with a common interest).

Steinfield, then a doctoral student in communication research at the University of Southern California (and now a professor at Michigan State), obtained the company's agreement to cooperate in his research. He became a member of two of the distribution lists (each consisted of several hundred individuals) in the electronic mail system. Such entre allowed Steinfield to participate in the interactive system, and to gain an initial conception of the functions that it served. His unobtrusive data-gathering allowed him to identify various uses of the system; he then included this use scale in the questionnaire that he distributed to a random sample of 400 users. A response of 55 percent (N = 220) was achieved.

Each respondent was asked how frequently (if at all) they used the electronic mail system for each of eighteen different uses (the use scale). A factor analysis of these data indicated two main dimensions of use: *task-*related, including such uses as sending a message in place of a phone call, scheduling meetings and appointments, and keeping a record of agreements and other matters; and *socio-emotional,* including playing games, keeping in touch/maintaining relationships, and learning about events of interest. Other socio-emotional uses of the electronic mail system included taking breaks from work, advertising, organizing social activities, and filling up free time. So the electronic mail system had important social and entertainment functions. It was used something like a magazine or company newsletter.

What independent variables were most highly related to the degree of use? One important determinant of use was access, such as by having one's own computer terminal, rather than having to share it with others. Another independent variable highly related to use was need, indicated by whether or not an individual had to communicate with co-workers in other divisions of the company or who were officed in other cities (the electronic mail system was especially advantageous in overcoming physical distance). Finally, various social characteristics of the individuals on the electronic mail system (such as age, office rank, and length of time as an employee) were related to their degree of use.

One of the unique uses of the electronic mail system (which was not possible via other communication media) was to broadcast requests for information. For example, a message might be sent to a large distribution list, beginning with "Anyone know anything about . . . ?" Another unique use occurred when a design team surveyed the user population for their reactions to several design alternatives. Within a day, the design team received several hundred responses. The team then fed back the survey results to the users, explaining that they followed the majority choice. No other communication technology could have allowed such rapid access to the company's physically scattered expertise on this issue. Generally, electronic mail allowed types of communication that otherwise could not take place without great difficulty.

Conventional wisdom about computer teleconferencing indicates that electronic mail does not work very well for such functions as conflict resolution or negotiation. Indeed, Steinfield (1983, p. 66) found that very little use was made of the electronic mail system for these purposes. The interactive communication system did not have enough social presence for handling conflict or bargaining.

Nor was electronic mail used much as a means of getting to know someone. Further, confidential matters were seldom sent over the electronic mail system, out of concern about the security of the system.

Innovation Clusters and the Hot Market

Compared to the rate of adoption of most other innovations, the new communication technologies are spreading very rapidly among American households. Perhaps there is a relatively small segment of the total population, say 5 to 10 percent, that consistently are the first to adopt each of the new communication technological innovations such as microcomputers, videotape cassette recorders, cable TV, videotext and teletext, and the like. These innovators evidently realize they are part of an Information Society, and place a high value on technological innovations that help them obtain information. In Japan, this 5 to 10 percent of households who are generally the first to adopt new communication technologies are called the "hot market"; they tend to purchase microelectronics innovations almost as soon as they become available to consumers.

Some evidence for the existence of a hot market of enthusiastic superinnovators in the United States is provided by a 1984 study by Mediamark Research, Inc., a New York market research firm. A sample of 20,000 adults provided data about their adoption of 300 consumer innovations. A hot market for new communication technologies was represented by individuals who had adopted a cluster of electronic innovations: home computer, video cassette recorder, videodisc, and who had purchased sophisticated stereo equipment for their home and car. These communication technology superinnovators, however, were not the same consumers who had joined health clubs, acquired cash management accounts, or installed woodburning stoves. Thus, innovativeness was specialized by broad classes of innovations, with the new communication technologies forming one of the clusters of consumer innovations.

If such innovation clusters exist, and the adoption of one new medium affects the adoption of others, are there certain "trigger innovations" that, once adopted, lead a consumer to purchase others? The answer to this research question has not yet been adequately explored, but it is an important lead for future research.

CHARACTERISTICS OF THE ADOPTERS OF THE NEW MEDIA

What characterizes the individuals and households who are the first to adopt new communication technologies? One of their most striking characteristics is their socioeconomic status. Whether measured by income, occupational prestige, or in years of formal education, innovative individuals are relatively more elite than those who adopt later (or who reject). Basic reasons for the innovativeness/status relationship are that (1) the new media represent a nontrivial cost, and socioeconomic elites are in a better position to be able to pay, (2) the more-educated are more likely to be aware of the importance of information and to feel a need for it, and (3) certain high-prestige occupations (for example, scientists and engineers) are key figures in the coming Information Society, and they are technically more competent to use the new communication technologies. The positive relationship between socioeconomic status and innovativeness is a basic reason why the new media are widening the information-gap in society between the information-rich and the information-poor (as we detail in Chapter 5). Because the information-rich adopt the new communication technologies relatively earlier than the information-poor, thanks in part to the higher socioeconomic status of the information-rich, these individuals, households, and organizations gain in information level more rapidly than do the information-poor.

Compared to the conventional mass media of radio, television, film, and perhaps the press, the new media have a higher ratio of information to entertainment content. Most of the computer-based communication systems (videotext, teletext, teleconferencing, and interactive cable TV) mainly provide information to their users. Cable TV and video cassette recorders are predominantly utilized for entertainment purposes, of course. Certainly the information-orientation of the new media is a factor in their adoption and use; individuals who wish to complement the heavy entertainment diet of the conventional media are attracted particularly to the new media. These individuals tend to be highly educated, especially those who are employed in such information work as R & D, education, and the professions (Dutton and others, 1986).

Higher-status individuals may be especially likely to adopt the new media because they are seen as status symbols. When the Terminals for Managers (TFM) service, an electronic messaging system, was introduced to the eighty top officials at Stanford University in 1980, it was widely perceived as a mark of social distinction to be invited to participate. Our personal interviews indicated that 58 percent of those using the system felt that having a TFM computer terminal on one's desk functioned as a status symbol (Rice and Case,

1983). The following year, 1981, TFM was expanded to include another hundred or so university administrators; certainly many of these individuals adopted TFM eagerly because they were thus joining an exclusive electronic club.

Earlier adopters also have different communication behavior than do later adopters. Earlier adopters are:

- More cosmopolite (*cosmopoliteness* is the degree to which an individual is oriented outside the social system)
- More exposed to mass media channels, and relatively less dependent on interpersonal communication channels
- More exposed to interpersonal communication channels, and more highly interconnected through network links to the system
- More directly in communication with scientific and technical sources of information about the new communication technologies.

So the general picture of the communication behavior of the more innovative individuals in adopting the new media is that they are relatively more active information-seekers about such innovations, that they pursue such information from more expert sources, and that they range far afield in their travel, reading, and friendships.

Finally, earlier adopters of new communication technologies differ from later adopters in certain personality variables. They have greater *empathy* (the ability of an individual to project himself or herself into the role of another person), less *dogmatism* (the degree to which an individual has a relatively closed belief system), a greater ability to deal with abstractions, and more *rationality* (the use of the most effective means to reach a given end).

Past research shows that earlier and later adopters of new communication technologies differ in socioeconomic status, communication behavior, and personality variables (Rogers, 1983, pp. 251–270). Perhaps there are two other personal characteristics of adopters that should be mentioned here: age and sex. Children are much more receptive to the new communication technologies than are adults, and are able to learn how to use them with much greater facility. Illustrations are microcomputers and video games. Individuals past middle age, especially females, encounter great difficulty in learning to use computers and other of the new media. In

Chapter 5 we will detail evidence that boys are more receptive than girls to the use of microcomputers.

The Rapid Diffusion of VCRs*

One of the new technologies that started off with a rapid rate of adoption during the 1980s and promises to climb the S-shaped diffusion curve even faster in the late 1980s is video cassette recorders (VCRs). Figure 4–4 shows that 20 percent of American households had purchased VCRs by 1985, up from about 1 percent in 1980.

Why have VCRs diffused so rapidly in the United States? One reason is a favorable 1984 Supreme Court ruling that home videotaping of television broadcasts is not a violation of copyright laws. This decision ended years of litigation, and ensured the clear right of VCR owners to copy anything directly from their television sets without being accused of

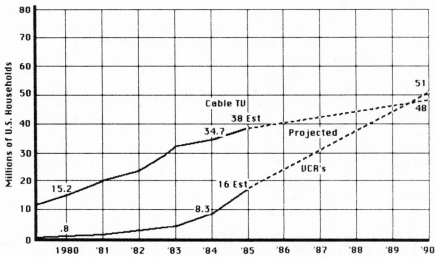

Figure 4–4. The Rate of Adoption of Video Cassette Recorders Is Increasing More Rapidly Than the Rate of Adoption of Cable TV

Among the approximately 80 million households in the United States, the rate of adoption for VCRs was less than half that for cable TV in 1985, but the VCR rate of adoption is increasing faster, and will pass the number of adopters for cable TV by 1990, according to projections. Why do you think the rate of adoption of VCRs is so rapid?

Source: Based on data provided by the A.C. Nielsen Company.

piracy. But this legal decision obviously could not have affected the rapid rate of adoption already underway from 1980 to 1984.

One main cause of the rapid adoption of VCRs in America traces to falling prices. In 1975, when VCRs were first introduced in the U.S., Sony's Betamax home recorder sold in combination with a television set for $2,200. By 1985, the price had dropped to $300; blank videotapes decreased in price from $20 to $5. In a pattern generally similar to that of television adoption in the 1950s, the first adopters of VCRs were definitely "upscale" (higher than average in income and better educated). By the mid-1980s, socioeconomic differences between adopters versus non-adopters of VCRs were fading as this communication technology became more widely adopted. There is a two-way relationship between the cost of VCRs and their rate of adoption; not only did their lower cost cause more rapid adoption, but the enlarging market also contributed to tougher price competition, and thus to lower prices.

Another reason for the fast take-off of the VCR diffusion curve came in the 1980s with a wide range of prerecorded cassette tapes becoming available for purchase or, more commonly, rental. Over 14,000 pre-recorded cassettes were on the market in 1985, including movies (about 7,000 are X-rated), music videos, children's programs, and various types of how-to cassettes (aerobic exercise, building a bookcase, etc.). It costs a family of four about $25 to see a recently released film at a movie theater; for a $5 rental fee, they can see the same film at home on their VCR.

What are the consequences of the fast diffusion of video cassette recorders in the United States? Big bucks for Hollywood film studios, who sell the movie rights for current film hits; such sales represented 14 percent of total film revenues in 1984. But the VCRs' widespread adoption has been bad news for network TV. The share of the viewing audience attracted by ABC, CBS, and NBC dropped from 92 percent in 1977 to 74 percent in 1984. Some of the former network TV viewers defected to cable television, but many are watching videotaped programming.

Naming a New Communication Technology

How important is the name of a new communication technology as a factor in its acceptance? The name determines, in part, the perception of the innovation by potential adopters, and thus affects whether they will adopt and use it. Often, the naming of a new medium is dominated by technical experts, and so the names come out sounding very technical (which may confuse or turn off many potential adopters). Sometimes the naming of a communication technology is a rather haphazard activity; very seldom is formative evaluation research conducted to identify the most appropriate name in the eyes of potential adopters.

A haphazardly selected name sometimes may work out better than a too-technical term. An illustration is the Green Thumb videotext system used by Kentucky farmers in the early 1980s. Two U.S. government experts in weather news originally called this system AGWEX to denoted that it provided agricultural weather information through the extension service. A secretary in the office of United States Senator Walter Huddleston, who strongly supported this new communication technology, began calling it the Green Thumb (Paisley, 1983). The name stuck, and had high impact with Kentucky farmers.

The French teletext system, officially called Antiope (for *L'Acquisition Numerique et Televisualisation d'Images*) after Antiope, daughter of the king of Thebes in Greek mythology, is widely confused by the public with the French (and English) word *antelope*. In this case a very literary name was confusing to the potential users.

When the naming of a new communication technology is done haphazardly or by technologists without the benefit of formative evaluation to guide their choice of words that would be meaningful to the public, such names then hinder acceptance of the new technologies by the public. New communication technologies ought to be given names that are meaningful and understandable to the users. Instead, names are often given in a way that, while technically correct, may confuse potential users or convey a negative impression. We ought to devote much more care than in the past to selecting the name for a new communication technology.

The Innovation Process in Organizations[4]

During the first several decades of research on the diffusion of innovations, these studies focused on individuals as the units of adoption. Then it was realized that many innovations are adopted not by individuals, but by organizations. This shift in the unit of adoption is particularly important in the case of the new communication technologies, which frequently are adopted by organizations: for instance, microcomputers in schools, and electronic mail and teleconferencing in business firms.

Studying the innovation process in organizational settings represents a shift from the classical diffusion model in several important ways. For example, the innovation process in organizations is usually considered successful if it leads to implementation (including institutionalization of the new idea), not just to the adoption decision per se. Also, as mentioned previously in this chapter, the individual plays an active, creative role in the innovation process by matching the innovation with a perceived organizational problem, and perhaps in reinventing the innovation. An innovation should not be conceived as a fixed, invariant, and static element in the innovation process, but as a flexible and adaptable idea that is consecutively defined and redefined as the innovation process gradually unfolds.

Finally, instead of the past overdependence on surveys and cross-sectional data analysis of the correlates of innovativeness, researchers of the innovation process in organizations began to engage in more in-depth case studies. The shift from highly structured, quantitative methods of investigating diffusion behavior led to more qualitative, hypothesis-generating case studies of the innovation process in organizations.

Certainly, innovation is a *process*, a sequence of decisions, events, and behavior changes over time. Past research designs did not adequately allow an adequate analysis of the temporal aspects of innovation necessary to explore its process nature. Very little past research included data at more than one observation point, and almost none at more than two such points. Therefore, almost all past research was unable to trace the change in a variable over "real" time; these past investigations dealt only with the present tense of innovation behavior.

A MODEL OF THE INNOVATION PROCESS

Recent investigations of the innovation process in organizations followed a relatively unstructured, open-ended, case-study approach to the gathering of data. They are "tracer studies" of the innovation-decision process for a single innovation in an organization. These investigations are guided by a model of the innovation process in organizations, which is conceptualized as consisting of five stages, each characterized by a particular type of information-seeking and decision-making behavior. Later stages in the innovation process usually cannot be undertaken until earlier stages have been settled. Thus, there is a logical sequence to the five stages in the innovation process, although there are exceptions. The five stages are organized under two broad subprocesses, initiation and implementation, divided by the point at which the decision to adopt is made (Figure 4–5).

Initiation

The first two stages of the innovation process are together known as the initiation phase or subprocess. *Initiation* is all of the

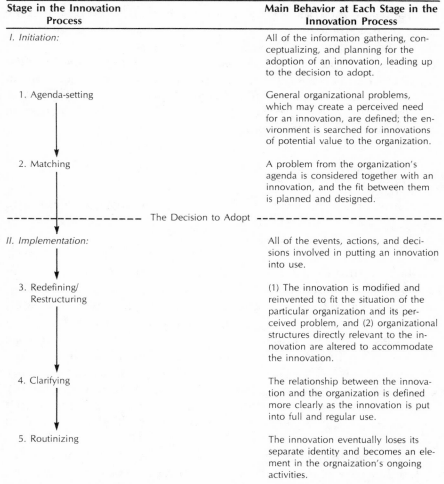

Stage in the Innovation Process	Main Behavior at Each Stage in the Innovation Process
I. Initiation:	All of the information gathering, conceptualizing, and planning for the adoption of an innovation, leading up to the decision to adopt.
1. Agenda-setting	General organizational problems, which may create a perceived need for an innovation, are defined; the environment is searched for innovations of potential value to the organization.
2. Matching	A problem from the organization's agenda is considered together with an innovation, and the fit between them is planned and designed.
————————— The Decision to Adopt ————————————————	
II. Implementation:	All of the events, actions, and decisions involved in putting an innovation into use.
3. Redefining/ Restructuring	(1) The innovation is modified and reinvented to fit the situation of the particular organization and its perceived problem, and (2) organizational structures directly relevant to the innovation are altered to accommodate the innovation.
4. Clarifying	The relationship between the innovation and the organization is defined more clearly as the innovation is put into full and regular use.
5. Routinizing	The innovation eventually loses its separate identity and becomes an element in the orgnaization's ongoing activities.

Figure 4–5.   A Model of Stages in the Innovation Process in Organizations

SOURCE: Rogers (1983). Used by permission.

information-gathering, conceptualizing, and planning for the adoption of an innovation, leading to the decision to adopt.

Stage #1: Agenda-setting. The first step in the initiation of an innovation occurs when an organization's leaders define a general organizational problem, which may create a perceived need for an innovation and set off a search of the organization's environment for innovations that may help solve the organization's problem. Agenda-setting is a continuous process in organizations that occurs when administrators look within the organization to identify problems and create solutions, and leaders seek answers outside the institution's boundaries. To a certain degree, innovations can create their own demand.

Stage #2: Matching. Leaders match the innovation with their organization's problem to determine whether it fits their needs. Typical of such an assessment is creation of a mental scenario in which the innovation is vicariously adopted and administrators estimate its impacts. At this point a decision is made whether to adopt, perhaps on a trial basis, or to reject.

Implementation

Three stages in the innovation process follow the decision to adopt. *Implementation* is all of the events, actions, and decisions involved in putting an innovation into use.

Stage #3: Redefining/Restructuring. The application of a new idea often differs from the uses planned before adoption. Thus, an adjustment must begin. If the fit between the problem and the innovation is good, then alterations are minimized. In many cases, however, at least some degree of modification occurs; this is re-invention. The innovation usually causes modifications, which are sometimes profound, in the structure of the organization.

Stage #4: Clarifying. As the innovation becomes integrated into the day-to-day operations of the organization, its meaning gradually becomes clear to the members of the organization. The innovation eventually loses its newness and becomes embedded in established protocols.

Stage #5: Routinizing. Now the innovation has been well-defined and becomes part of the infrastructure of the organization. Eventually, the innovation loses its newness and is no longer recognized as a separate entity in the organization. It has just become part of the ongoing procedures of daily operations.

The Diffusion of Microcomputers in California High Schools*

Now we draw on the results of an investigation of the adoption of microcomputers in nine California high schools (Rogers and others, 1985b and 1985c) in order to illustrate the nature of the innovation process. Figure 4–6 shows the five-step innovation process for microcomputers at Milpitas High School, a California school system with an enrollment of 1,700 students, 42 percent of whom are minority members (principally Hispanic and black). In 1984, there were thirty microcomputers in the school, plus two minicomputers. The main events in the innovation process seem to generally fit fairly well into the first four of the five stages in the model (routinizing had not yet occurred at the time of our data gathering).

Stage #1: *Agenda-setting*. In 1979, the high school's assistant principal became aware of the changing demographics of the Milpitas community as blue-collar workers were being replaced by white-collar professionals employed in the nearby Silicon Valley electronics companies. This change sparked his idea of adding microcomputers to the school's educational tools as more of the Milpitas students were seeking preparation for four-year college. The adoption of microcomputers became one part of a more general plan to update and enhance the high school's curriculum. The assistant principal also had an administrative reason for introducing microcomputers: The machines allowed him to employ less-skilled employees in the school's administrative office.

Stage #2: *Matching*. By 1979, the assistant principal had decided that the computer was an important educational asset, but he was unsure how it might offer solutions to match Milpitas High School's academic needs. He sought technical assistance by hiring Bob Lennon (a pseudonym), a math/physics teacher who had been laid off by a nearby school district. Lennon became the "computer champion" at Milpitas High School.

During the 1980 school year, the assistant principal and Lennon traveled to a nearby high school to observe its use of microcomputers. They discovered that the school's students kept the machines busy from when the school's doors opened in the morning until they closed at 6:00 P.M. Lennon and his administrator were much impressed by this sense of excitement about learning. "I don't give a damn whether you're learning to tie your shoes," said the assistant principal. "The important thing is to get turned on to the process of learning."

Stage #3: *Redefining/Restructuring*. Milpitas High School's computer experience began with seven terminals connected via dedicated telephone lines to the school district's central computer. The first computer classes were taught in 1982 (which was relatively late compared to neighboring schools). Since then the school added thirty microcomputers, and in 1983, the school received two donated DEC PDP-1145 minicomputers (which school officials estimated to be worth $200,000).

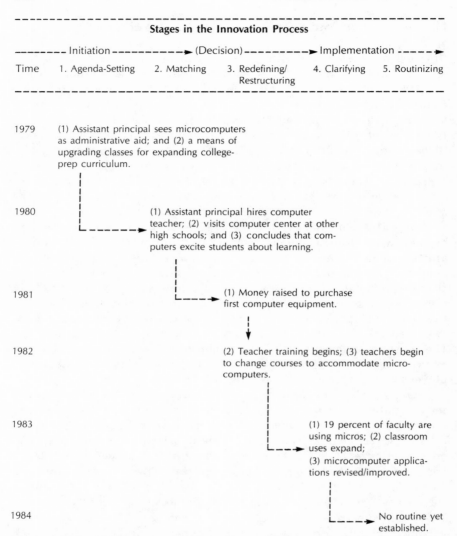

Figure 4-6. Classifying the Adoption of Microcomputers at Milpitas High School by Stages in the Innovation Process

SOURCE: Rogers and others (1985b). Used by permission.

Milpitas High put the new microcomputers in several locations: fourteen in a computer center, two in the school's business department, two in the science department, three in special education, one in industrial arts, and eight in remedial classrooms. About fifteen Milpitas teachers redefined their classroom protocols to include the computer, of these, five teachers offer mathematics using the new machines, two business, two special education, and two science; four adopting teachers are in various other disciplines.

Stage #4: Clarification. Microcomputer adoption by the end of the 1983–1984 school year had reached 19 percent of the school faculty of eighty teachers, a remarkable rate of diffusion in a relatively short period of time. Among these teachers, clarification had begun, with microcomputer uses and programs being defined and generally expanded.

Why did the innovation process proceed so quickly at Milpitas High School? First, championship of the innovation originated with the assistant principal, an individual near the top of the school's hierarchy. One teacher, Bob Lennon, was assigned the sole task of becoming the in-house technical expert. He provided computer training programs for a number of teachers. Funding was available for purchasing the thirty microcomputers through a state grant for school improvement, from school district special funds, and from corporate donations. The decentralized positioning of the microcomputer hardware made it accessible to teachers in disciplines other than just business and mathematics, the most common subjects in which microcomputers are used in most high schools.

Stage #5: Routinization. For Milpitas, and for most of the other eight high schools in our study, talk of "routine" uses of microcomputers was still premature in 1985. Only computer programming classes, which enrolled close to 300 of Milpitas's 1,700 students, might be considered a stable application of microcomputers. But even here changes are reported, such as an expansion from BASIC into more sophisticated programming languages such as Pascal.

Important Actors in the Innovation Process

Innovation is a keenly social process, so it is important to examine the key social roles that govern the speed and adequacy of implementation.

1. The "computer champion" is the spark plug of the innovation process. This individual is typically a teacher of mathematics, business, or science who saw the usefulness of computers in instruction before others in the high school. This person—and it usually is only one person at the start of the innovation process—begins the diffusion of computers through conversations with other teachers and administrators. The computer champion may be a present member of the faculty or someone hired specifically for his or her computer expertise.

Interestingly, the computer champion often does not remain at one high school, or at least does not stay in his or her previous full-time instructional capacity. In two of the nine California high schools that we studied, the computer champions were tapped by the school district for administrative work in diffusing computers throughout the school district. In another case, the champion began writing instructional software in what became a transition out of the classroom into self-employment as an entrepreneur in educational computing.

In many respects, the computer champion fits the role of the innovator in past diffusion studies. He or she may be considered a little odd by co-workers and is likely to have a cosmopolite circle of friends that reaches beyond fellow teachers in the local school. The innovator must be a risk taker, one who rates his or her own efficacy highly. Such high self-efficacy is not the norm among schoolteachers.

2. A second key actor is the school administrator. If the computer champion was the principal or closely allied with the principal, the spread of microcomputers for instruction in a California high school was markedly faster and more coordinated than when the champion was a teacher working without administrative support.

Although we found no principals who said they were hostile to instructional computing, the degree of their enthusiasm for, and familiarity with, computers was a major determinant of the extent to which microcomputers were integrated into a high school's curriculum. An obvious reason for this pattern is the principal's ability to direct budget and personnel toward microcomputer applications.

3. A third key role is played by the parents of schoolchildren. Through direct contact with teachers and administrators, as well as in parent-teacher organizations, parents influence a school's direction regarding microcomputers. They raise money for computer purchase, pressure school administrators, and generally have a very direct stake in this new technology.

4. The fourth actor in microcomputer diffusion is the computer industry. Many computer firms have national gift programs for schools. Apple Computer Inc., in 1983, donated one computer to each school in California. Computer companies are generally eager to donate equipment to schools, or at least to provide heavy price discounts. The industry's generosity is not without purpose. Computer manufacturers have learned that brand loyalty transfers from the school to the home or corporate purchase, or to the next school purchase. Donated equipment also represents a corporate tax write-off.

Schools have found that "free" computers actually can be quite expensive. Software and peripherals such as printers and disk drives must be purchased before schools can fully operationalize their systems. A gift computer can become a Trojan Horse for a high school after the time and cost of setting up a computer center, training teachers, maintaining the system, and updating courseware have been encountered. One donated minicomputer required about $300-a-month preventive maintenance. "One more gift like that," the school administrator said, "and we'll go under."

Although a pioneering teacher may take the lead in introducing computers in a high school, administrators usually must become involved in implementing the technology because organizational resources are required for adoption and because reward structures must encourage teachers to adopt. Microcomputer applications reached across more school subjects and have involved more students in California high schools where the principal participated early in the innovation process.

The Degree of Microcomputer Implementation

Before 1982 there were only a handful of microcomputers in the nine California high schools we studied. Most were used for administrative tasks such as keeping students' grades and attendance records. The number of students involved in classes using computers was seldom more than 1 percent of the enrollment. Very few teachers had ever used a computer.

By 1984, these high schools had adopted 225 microcomputers, a tenfold increase in two years. About 120 more microcomputers were added in 1985. However, there was still only one microcomputer for every sixty students. And the number of teachers expert in the command of computers and programming remained very small, as few as one or two per school (our nine schools of study each had from fifty to eighty teachers). But the number of teachers who had beginning skill levels — say, adequate to teach word-processing — grew dramatically from 1982 to 1984. Teacher workshops for one or several days were sponsored by the school district, summer courses on computers were featured, introductory classes were offered by computer companies and individual study (often accompanied by the purchase of a microcomputer for the teacher's home) all helped drive this surge in computer knowledge. Estimates of minimum "computer literacy" among the high school teachers in our study range from 10 to 50 percent of the total faculty.

These two-tiered levels of microcomputer expertise — a very small number of teachers with strong computer skills and a moderate number with slight familiarity with microcomputers — is the direct result of budget constraints, and, to a lesser extent, of the decentralized nature of this innovation's diffusion. Funding for microcomputer implementation, including teacher training, is outside the regular school budget and consists of one-time grants and donations, such as from computer companies.

As a result of financial restrictions, the greatest obstacle to the further diffusion of microcomputers in high schools is imposed by the school's infrastructure (Figure 4-7). The usual person-to-person spread of an innovation is hampered by the lack of a reward structure in schools encouraging adoption. Teachers seeking an in-depth knowledge of computing have usually had to learn it on their own. In order for the rate of diffusion of microcomputers to "take-off" so as to reach an adoption level of around

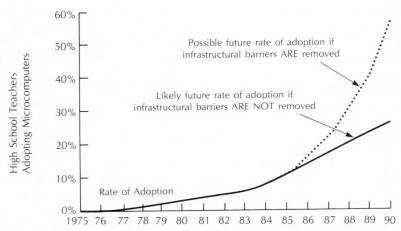

Figure 4-7. Likely and Possible Future Rates of Adoption for Microcomputers in High Schools, Depending on the Removal of Infrastructural Barriers

In the nine California high schools studied, the rate of adoption of microcomputers got off to a rapid start in 1983–1985, reaching about forty computers per school. However, this amounted to only one microcomputer for every sixty students, and only one or two teachers (out of a school faculty of fifty to eighty) were expert enough to create their own computer programs. A massive investment of resources would be needed to get the S-shaped diffusion curve to take off.

50 percent, considerable additional resources will be needed. Certain teachers would need a year or more of release time from regular teaching duties to create appropriate microcomputer software and to integrate such instructional materials into their courses. Many such teachers would need to collaborate with computer programmers, instructional designers, and subject-matter specialists in order to create the needed software. Funding for such a large-scale effort is certainly not currently available in our California high schools of study. This lack represents the major infrastructural barrier to further adoption of microcomputers in these schools. Figure 4-7 illustrates the nature of infrastructural barriers to effective diffusion of microcomputers in high schools. Without a major push toward creating appropriate easy-to-use software, followed by a massive training program for teachers, the S-curve of microcomputer adoption will not take-off.

Perhaps microcomputers will help instill renewed excitement and confidence in public schooling to the point where each student will have an electronic tutor patiently and individually pacing instruction, and where teachers are freed from the drudgery of their profession to inspire and shape young minds. That visionary potential is yet far from being achieved in the nine California high schools of our study.

Summary

The adoption and implementation of the new communication technologies can best be studied and understood in light of the theory of the diffusion of innovations. The main elements in the diffusion of new ideas are: (1) an *innovation* (defined as an idea, practice, or object perceived as new by an individual or other unit of adoption) (2) that is communicated through certain channels (3) over time (4) among the members of a social system. Three distinctive aspects of the diffusion of the new media are their (1) critical mass nature (which means that each successive adopter increases the value of a new communication technology for all previous adopters), (2) *re-invention* (the degree to which an innovation is changed or modified by a user in the process of its adoption and implementation, which frequently occurs because the new media are tool technologies that can be applied in a variety of ways), and (3) the focus on implementation and use, rather than just on the decision to adopt.

Several of the new media diffused very rapidly during the 1980s: microcomputers, VCRs, and cable television. On the other hand, the diffusion rate for ATMs has plateaued at one-third of the American public; perhaps for the two-thirds who are rejectors of this innovation, the socio-emotional relationship with one's banker cannot be replaced by a task-efficient computerized teller. Similarly, the rate of adoption of teletext and videotext, both in the United States and in other nations, has been discouraging (Chapter 2).

Earlier adopters of the new media, compared with later adopters, have higher socioeconomic status, different communication behavior (for example, they are more cosmopolite and more likely to utilize mass media channels), and have greater empathy, less dogmatism, and more rationality. Further, younger age and male gender is associated with earlier adoption of the new media, especially computers.

Many of the new communication technologies are adopted and implemented by organizations, not by individuals. Illustrations are microcomputers in schools and office automation in companies. Here we described a five-staged model of the innovation process in

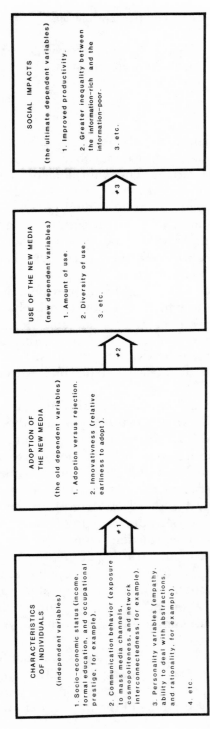

Figure 4–8. The Main Types of Variables in Research on the Adoption, Use, and Social Impacts of the New Communication Technologies.

Most research following the classical model of the diffusion of innovations has looked only at the relationships depicted above as arrow #1 (variables related to adoption). Study of the new communication technologies has broadened diffusion research to also include investigation of the relationships identified as arrow #2 (indicating the correlates of use), and as arrow #3 (factors related to the social impacts of the new media). This latter topic is discussed in Chapter 5.

SOURCES: Based on Rogers (1983, p. 376), Dutton and others (1985), and Dutton and others (1986).

organizations: (1) agenda-setting, (2) matching, (3) redefining/restructuring, (4) clarifying, and (5) routinizing.

Figure 4-8 summarizes the big picture of this chapter, and shows how adoption and use relate to the social impacts of the new communication technologies.

Research on the diffusion of innovations has been given a boost in recent years by the studies of the adoption and use of the new communication technologies.

Notes

1. A *dependent variable* is the main variable in which an investigator is interested and seeks to explain in a research study.
2. Similar results were obtained in a survey of Canadian home computer diffusion; 57 percent of the users reported watching less television. They also reported a decrease in time spent with books, magazines, newspapers, and radio (Caron and others, 1985). Vitalari and others (1985) found a similar reduction in television viewing among their California home computer adopters.
3. This case study draws on material from Daniel Hertzberg (February 21, 1985), "If Carrots Don't Persuade People to Use ATM's, Banks Go for Sticks," *Wall Street Journal.*
4. Certain of the ideas in the following section are adapted from Everett M. Rogers and others (1985a and 1985b) and Rogers (in press).

CHAPTER 5

Social Impacts of Communication Technologies

"The new media are shaking the foundations of how communication can occur. Designations such as 'mass media,' 'interactive,' 'broadcast,' and even 'communications' are being redefined on-the-fly. The scale of impacts is staggering. The potential uses are intriguing. The impacts are puzzling."

Robert Johansen, 1984

"For whosoever hath, to him shall be given, and he shall have more abundance: but whosoever hath not, from him shall be taken away even that he hath."

Matthew, 13:12, *The Holy Bible: King James Version*, New York, Cambridge University Press

In the present chapter, we shall explore what is known about the social impacts of the new communication technologies. This current work derives from a background of earlier studies of audiences and media effects. Then we look at the various social impacts of the interactive communication technologies on individuals, organizations, and society.

From Audience Research to Effects Research

Today it may be difficult for mass communication researchers in North America and Western Europe to remember that our field began with a primary focus on determining such relatively simple

150

issues as the size of media audiences and the characteristics of the audience members. Describing the audience is a natural first question for communication researchers to ask. Only later did communication scholars gain interest in more theoretical questions, such as what effects the mass media have on their audiences, the processes by which such effects occur, and the personal and social functions that the media serve (Rogers and others, 1985a).

In the United States, over the years, as each new communication technology was introduced, the first scholarly research on it mostly consisted of audience surveys (Chapter 3). For instance, George Gallup (1930) and Ralph Nafziger (1930) reported the first audience studies of newspapers; their basic research paradigm comparing readers to nonreaders has continued to this day, with more than fifty studies of U.S. newspaper published from 1950 to 1977 (McCombs, 1977), and with another fifty or so published since then. During the past thirty-five years, however, primary scientific attention has been given to investigating the *effects* of newspapers (and other mass media) on readers' awareness, attitudes, and actions.

In a parallel development, radio research in the United States began with audience surveys, such as those by Paul F. Lazarsfeld and Frank Stanton (1941 and 1944), and then shifted to investigations of radio effects. Similarly, the earliest communication research on film surveyed audiences, especially children, and then moved to a focus on effects, particularly potentially negative effects (for example, the Payne Fund studies of film effects on children reported by W. W. Charters [1934]). Later, during the rapid growth of television ownership in the United States during the 1950s, audience surveys of television viewers soon gave way to effects-oriented studies by communication scholars, especially of children (Schramm and others, 1961; Himmelweit and others, 1958).

Past Research on Communication Effects

Just as we saw, in the previous chapter, that adoption/implementation research on the new media had a basis in past research on the diffusion of innovations, so the study of the social impacts of the new communication technologies had a basis in past research, this time on communication effects. What are communication effects? *Effects* are the changes in an individual's behavior (knowledge, attitudes, or actions) that occur as the result of transmission of a communication message. The study of effects generally assumes a linear model of communication; the ability to infer effects is equivalent to the ability to infer causation (the impact of a presumed cause on a presumed effect). Determining causation is a very tricky business in any scientific field, including communication research. Perhaps it is impossible, at least in an absolute sense. We can only draw inferences, rather than definite conclusions, about the cause of an effect. In any event, as we have already stressed, the most popular topic of research by communication scientists has been the investigation of effects, both from the mass media and from interpersonal communication. Unfortunately, because the new media are interactive and are also different in certain other ways, much of the effects research paradigm cannot be applied to the contemporary studies of the social impacts of the new communication technologies. Nevertheless, we need to review the communication effects research here, as background to our discussion of the social impacts of the interactive communication technologies. Impacts research grew out of effects research, and is much influenced by it.

The past forty years of research on communication effects has followed a series of ups and downs, an alternating optimism and pessimism about the degree to which the mass media have powerful effects. Here we briefly trace this cyclical history.

THE ERA OF MINIMAL EFFECTS

The minimal effects era reached its peak during the 1940s and 1950s in the United States, crystallized by a series of investigations mounted by communication scientists at the Bureau of Applied Social Research, Columbia University, under the direction of Paul F. Lazarsfeld, one of the four founding fathers of communication research (Chapter 3). Lazarsfeld and his colleagues described their research results on the limited effects of mass communication in counterpoint to the *hypodermic needle model*, which stated that the mass media have direct, immediate, and powerful effects on their audience. Today, we cannot find any clear statement of this model by a responsible communication scholar prior to the minimal effects era, or any research-based evidence for its existence, and it seems that Lazarsfeld and his co-workers set up this model as a strawman to destroy with the limited effects findings from their survey research studies.

Whether or not academic communication scholars actually stated the hypodermic needle model, the notion was certainly widely held by the American public in the past. The reasons for believing that the mass media were very powerful centered on certain historical events:

- The role of the Hearst newspapers in promoting the shibboleth, "Remember the Maine," as a means of building public support for U.S. involvement in the Spanish-American War
- Orson Welles's "War of the Worlds" radio broadcast in 1938, which panicked about 30 percent of its American audience
- Nazi Germany's use of radio propaganda as a weapon during World War II
- The influence of Madison Avenue advertising in marketing products to unwary consumers, a concern voiced by pop writer Vance Packard in his 1950s best-seller, *The Hidden Persuaders.*

Eventually, when communication research was conducted on mass media affects, starting in the 1940s, considerable doubt was cast on the hypodermic needle model. It was based mainly on intuitive theorizing about untypical historical events, and was too simple and too mechanistic to accurately describe the effects of the mass media.

An early and influential contribution to the minimal effects model of the mass media was the research project by Lazarsfeld and others on the role of the media in the 1940 presidential election. Lazarsfeld and his investigators from the Bureau of Applied Social Research conducted personal interviews with a panel of citizens in Erie County, Ohio, in order to determine the media's influence in changing voting decisions. In this case, the media were found to have little direct effect in the election campaign (Lazarsfeld and others, 1944). The Erie County study, more than any other investigation, launched the research tradition of the minimal effects of the media. This view considered the mass media as just one among various other influences on individuals' behavior, and not a very important influence at that.

Other Bureau of Applied Social Research projects were carried out by Lazarsfeld and his staff during the 1940s and published in the 1950s. The media were generally found to be much less influential on consumer decisions than were interpersonal networks. The strongest statement of the limited effects viewpoint was Joseph Klapper's *The Effects of Mass Communication* (1960), a synthesis of various media effects studies. His conclusion was a kind of ultimate piece of pessimism about media effects: "Mass communication *ordinarily* does not serve as a necessary and sufficient cause of audience effects, but rather functions among and through a nexus of mediating factors and influences. These mediating factors are such that they typically render mass communication a contributory agent, but not the sole cause, in a process of reinforcing the existing conditions" (Klapper, 1960, p. 8).

Two mass media campaign studies that well represent the stamp of the minimal effects era were Herbert Hyman and Paul Sheatsley (1947) and Shirley Star and Helen Hughes (1950). The Hyman/Sheatsley article, entitled "Some Reasons Why Information Campaigns Fail," reviewed several survey evaluations of information campaigns that were designed to increase knowledge of international news and public affairs issues in the United States. The body of minimal effects literature that was reviewed led to the conclusion that "those responsible for information campaigns cannot rely simply on 'increasing the flow' to spread their information directly," such as by disseminating mass media messages. Several psychological barriers were identified by Hyman and Sheatsley as important limitations to campaign effectiveness: *selective exposure* (the tendency of an individual to attend to communication messages that are consistent with the individual's prior attitudes and experiences), *selective perception* (the tendency to interpret communication messages in terms of prior attitudes and experience), and *selective retention* (the tendency of an individual to remember communication messages that are consistent with prior attitudes and experiences). Hyman and Sheatsley referred to the hardest-to-reach audience members as "chronic know-nothings." Their choice of terminology obviously implied a pro-campaign bias, and a tendency to blame the audience for not being affected by a campaign's media messages.

The other landmark study, that by Star and Hughes (1950), reported a pre-post evaluation of the Cincinnati Plan for the United Nations. This six-month campaign sought to make residents of Cincinnati "United Nations–conscious." The campaign was very intensive, with 60,000 pieces of literature distributed, 2,800 organizations addressed by speakers, and radio stations airing up to 150 spots per week. Before-after surveys of about 800 adults by the University of Chicago's National Opinion Research Center indicated who was reached by the campaign's messages and with what effect. The pre-survey served as a formative evaluation[1] for the campaign, and led to the recommendation that the campaign be aimed particularly at women, the relatively uneducated, the elderly, and the poor (subaudiences who were poorly informed about the United Nations). However, the post-survey conducted six months later showed that the individuals actually reached by the campaign were the "better-educated, the younger, and the men." For the entire sample, only very minor cognitive effects of the campaign were detected; for instance, those individuals knowing at least something about the main purpose of the United Nations changed from 70 percent to only 72 percent during the six months of campaign activities (Rogers and Storey, in press).

A major shift occurred in public opinion during the campaign period, with the percentage of Cincinnati adults rating international problems (such as the likelihood of a third world war) as the most important issue facing the U.S., increasing from 56 to 77 percent. However, this change was almost entirely due to world news events, such as the increasing Cold War belligerence between the United States and Russia and without any of the shift in opinion actually due to the media campaign for the United Nations. Star and Hughes concluded: "Information grows interesting when it is functional, that is, when it is so presented that it is seen to impinge upon one's personal concerns" (Star and Hughes, 1950). The campaign slogan, "Peace Begins with the United Nations—and the United Nations Begins with You," illustrates the rather abstract, general nature of the Cincinnati campaign; the hoped-for increase in knowledge about the United Nations was not tied to any specific behavioral action, in particular a behavior that would impinge on the individual's own affairs. An example is a female respondent who, when questioned by a University of Chicago interviewer about the campaign slogan, replied: "Why, yes, I heard it over and over again. . . . But I never did find out what it means" (Star and Hughes, 1950).

Communication researches such as the Cincinnati campaign evaluation helped convince scholars during the minimal effects era that the mass media were not nearly so powerful as previously thought (and certainly not as impactful as the mythical hypodermic needle model implied). These campaigns mainly depended upon mass media channels to have effects, with relatively little attention being given to the role of interpersonal networks in causing campaign effects. Nor were such networks appropriately investigated in the research designs and data-gathering methodologies utilized in the minimal effects era. Only belatedly was it recognized that the intended audience for a campaign was not just a passive "target" waiting to be acted upon by the mass media; this notion is reflected in Raymond Bauer's article, entitled "The Obstinate Audience" (1964).

THE ERA OF CONDITIONAL EFFECTS

The minimal effects era for campaigns, and for the broader field of mass communication research, was followed by a period during the 1960s and 1970s in which it was realized that campaigns could succeed if they were designed and conducted according to certain communication strategies. The key document expressing the central theme of the more optimistic 1960s/1970s media campaign era is Harold Mendelsohn's "Some Reasons Why Information Campaigns Can Succeed" (1973). This author drew on three communication campaign experiences in which he had been involved, each considered successful, in order to identify such campaign strategies as utilizing formative evaluation in designing a campaign, so that accumulated communication research–based knowledge is incorporated, along with "feedforward" data about the campaign's audience. This formative evaluation strategy puts communication scientists in league with communication practitioners, rather than creating a tension between them, as when communication scholars simply tell practitioners (on a post hoc basis) that their campaign failed. "Most evidence on the failures of information campaigns actually tells us more about flaws in the communicator—the originator of messages—than it does about shortcomings either in the content or in the audience" (Mendelsohn, 1973).

Somewhat greater optimism about the degree of mass media effects also came from recognizing that the media had indirect effects, as well as direct effects. Similar to the way we define the concepts of direct and indirect impacts later in this chapter, *direct effects* are the changes in an individual's behavior that occur in immediate response to the transmission of a communication message, while *indirect effects* are the changes in an individual's behavior that result from the direct effects of a communication message. During the 1970s, communication scholars began to investigate a variety of indirect media effects. One of these was time-displacement; with a television set on for seven hours a day in the average American household, much less time is usually spent reading books and other written material, to playing games and other types of family entertainment, etc. Another indirect effect of the mass media is agenda-setting, the process through which the mass media communicate to an audience the relative salience of various news items. While the media may not be very effective in telling us what to think, they are often very effective in telling us what to think about.

Many of the indirect effects of the mass media are not clearly recognized by individuals at the time they occur, making them all the more powerful. When we are influenced but do not know it, the influence is especially strong. For instance, few of us are fully aware of the agenda-setting function of the media (until we retrospectively analyze how we become interested in certain news topics). So the indirect media effects may be even more important than the direct effects, even though they are not so visible. Another type of invisible media effect occurs when the mass media reinforce certain behavior. Perhaps an individual's behavior would have changed had the media not reinforced a contrary position. Is such nonchange in behavior a media effect or not? Early communication researchers during the era of minimal effects did not recognize it as such, but during the later era of conditional effects such behavior-reinforcing as occurred from selective exposure, selective perception, and selective retention, came to be seen as a certain type of media effects. But such reinforcement was difficult to measure in communication inquiry; change is a more visible marker than nonchange.

So by reconceptualizing the nature of mass media effects in terms of indirect effects and reinforcement, communication scholars since the mid-1970s have begun to detect certain media effects, at least more than were found during the preceding era of minimal effects. The post-1970s era is often called "conditional effects" because communication scholars were finding that the media had effects only under certain conditions, so that *some* media had *some* effects under *some* conditions on *some* people. An example is television violence effects on children; of the thousands of communication researches on this topic, most indicating that individual aggression is a consequence of TV violence, it is still not possible to say why some children are affected more and other children less. The main conclusion from the conditional effects era is that the power of the media are highly conditional, depending on a variety of contingent variables (Roberts and Maccoby, 1985). Television violence may be a catalyst that triggers children's aggressive behavior in certain cases, but it seldom is a direct cause.

Effects studies of the one-to-many media will undoubtedly continue to be carried out in future years by communication scholars. They will explore further the conditions determining the degree to which effects occur. Perhaps new models and new eras in media effects will emerge. More likely, a considerable portion of the research attention given to mass media effects in the past will be siphoned off to the study of the social impacts of the new communication technologies. That would be consistent with the pattern of the past, with each new communication technology as it came on the public scene in the United States soon gaining the main attention of communication scientists (Chapter 3).

PROCESS VERSUS EFFECTS IN COMMUNICATION RESEARCH

Since its inception, the field of human communication has maintained an intellectual commitment to the study of process. Almost every definition of the concept begins: "Communication is a *process* of. . . ." Despite this fact, the major research strategies employed by communication researchers have centered on the study of *effects*, and have largely ignored the study of process. More attention should be given to the pattern of communication effects over time, thus allowing investigation of the process of communication, rather than single-point-in-time effects. A poverty of the field of communication research is the lack of over-time data about communication effects.

The differences between process and effects notions of communication trace to the intellectual origins of communication theory, when it took off in the late 1940s with the publication of Norbert Weiner's *Cybernetics* (1948), and Claude E. Shannon's articles in the *Bell System Technical Journal* (1948). As we said in Chapter 3, the portion of Shannon's theory utilized by communication scholars was a linear, left-to-right, one-way model of communication (which included the concept of noise, but said virtually nothing about feedback in communication systems).

Most communication scientists were unwittingly mislead by Weaver's introductory chapter in Shannon and Weaver's (1949) book, and uncritically accepted the linear view of communication that Weaver implied. Since dynamic over-time data were virtually unavailable, and since social scientists were already familiar with statistical methods for static data analysis, they opted for single-point-in-time data designs that could be analyzed with such linear statistical methods as analysis of regression and analysis of variance. Thus, communication researchers focused primarily on the study of single-point-in-time effects, largely ignoring the over-time process or pattern-of-effects aspect of communication (Monge and Rogers, 1985).

These two contrasting theoretical issues (linear versus nonlinear models, and single-point-in-time designs versus over-time process research) are reflected in the title of Wilbur Schramm's influential book, *The Process and Effects of Mass Communication* (1954). The first chapter is an excellent conceptual discussion of communication-as-process, while most of the remaining chapters report the results of empirical research on communication effects. Five years later, the title of Joseph T. Klapper's book, *The Effects of Mass Communication*, indicated the main direction in which the field of communication research was moving. This main focus on communication effects has continued until the present, although it has come under increasing criticism in recent years as various alternative research approaches have been suggested (for example, Rogers and Kincaid, 1981; Rice and Rogers, 1984).

The research tools used by most communication scholars fit well with the study of effects: statistical methods derived from agricultural research on the effects of fertilizer and nutrition on plant and animal growth, one-shot survey data gathering, and computer data analysis. The research designs, and the research philosophies of early communication research had a heavy psychological origin (contributed by such founders as Carl Hovland and Kurt Lewin). The first two doctoral programs in communication research (at Illinois and Stanford) were both strongly connected with the psychology departments at those universities. Thus the research orientations of early communication scholars was particularly psychological; looking within individuals for explanations of their behavior, as psychologists tend to do, fit naturally with the study of media effects. The individual was the unit of analysis, rather than the individual's communication relationship with another.

Research can be classified into variance versus process categories (Mohr, 1982). Variance-type research designs are not capable of revealing the kind of process that many communication theories describe. The effects "model of explanation in social science has a close affinity to statistics" (Mohr, 1982, p. 42) and because statistical methods for testing hypotheses are so widely utilized in social research, variance approaches are much more frequent than are process approaches, even when the research problem of central interest calls for process research.

Most of our present-day statistical methods began in agricultural experimentation with the work of Englishman R. A. Fisher in the 1920s. As Gene Glass and others observed: "The Fisherian design which has so captured the attention of social and behavioral scientists was originally developed for use in evaluating agricultural field trials. The methodology was appropriate for comparing two or more agricultural methods with respect to their relative yields. The yields were crops which were harvested when they grew ripe; it was irrelevant in this application whether the crops grew slowly or rapidly or whether they rotted six months after harvest. For social systems, there are no planting and harvesting times. Interventions into societies and institutions do not have merely 'an effect' but 'an effect pattern' across time" (Glass and others, 1975, pp. 4–5).

Analysis of variance and other types of difference statistics are most appropriate for the examination of effects, but not for the examination of processes. While difference statistics may be able to tell us whether or not some process has occurred, such variance research seldom provides much understanding about *how* or *why* a process happens. Variance research is entirely appropriate for investigating certain research problems. But it cannot probe backward and/or forward in time to understand what happened first, next, and so on, and how each such event influenced the next. "In variance theory, time-ordering among the contributory (independent) variables is immaterial to the outcome" (Mohr, 1982, p. 43). Consequently, the static nature of variance approaches, and most of the statistical methods utilized in past communication research, limited understanding of communication as a process.

Such one-to-many media as radio, television, and film of the past fifty years in the United States, led to asking the important intellectual question of what their effects were. Effects questions were treated as important by political scientists (the effects of propaganda and of election campaigns, for example), by sociologists, and by psychologists, the social science disciplines represented by the four founders of communication science. So there are theoretical, historical, and technological reasons for the emphasis of past communication research on effects.

Mass media institutions and the professionals who worked in them had a strong pragmatic interest in media effects, and this practical interest was conveyed to communication scientists in terms of research grants for investigating effects. Many communication scholars came to research and teaching positions on a university faculty from a background of working in the mass media. Naturally, they brought with them their concern about media effects. Further, certain government agencies and private foundations (the Payne Fund, the Ford Foundation, the Markle Foundation, for example) have been concerned about the possible negative effects of the mass media, such as television violence on children. Out of this concern has grown funding for communication research on these potential media effects, in order to determine just how serious the negative consequences may be. So perceived social problems such as the effects of TV violence on children and the manipulation of voting choices by candidates' media campaigns, led to government and foundation funding for effects studies, and this funding turned the head of communication researchers.

So three different forces combined to head communication scholars in the direction of studying effects: (1) the Shannon and Weaver model of communication, which communication scientists interpreted as a linear model; (2) the one-way nature of the mass media of the day (newspapers and the electronic media of radio, film, and especially television); and (3) statistical methods that, because of their origins in agricultural experimentation, functioned best for investigating communication effects, and fit poorly with the study of human communication as process.

One of the major research tasks of communication scientists today is to understand the process and impacts of the Communication Revolution. This research direction is macro in scope, in comparison to such microscopic problems as communication effects that have captivated the past efforts of communication scholars. Such a major shift in focus as that from effects to impacts requires very major changes in communication theories and research methods. It demands an epistimological revolution.

A Typology of Impacts

In our analysis of social impacts, we find it useful to follow a threefold typology. Here we use *impacts* and *consequences* interchangeably, to mean the changes that occur to an individual or to a social system as a result of the adoption or rejection of an innovation.

1. *Desirable impacts* are the functional effects of an innovation on an individual or social system. A desirable impact helps an individual or system function more effectively. *Undesirable impacts* are an innovation's dysfunctional effects on an individual or social system. The terms "desirable" and "undesirable" are roughly equivalent to positive and negative in this context.

2. *Direct impacts* are the changes in an individual or social system that occur in immediate response to an innovation. *Indirect impacts* are the changes that result from the direct impacts of an innovation. In other words, indirect consequences are the direct effects of direct effects.

3. *Anticipated impacts* are changes caused by an innovation that are recognized and intended by the members of a social system. *Unanticipated impacts* are changes that are neither intended nor recognized. For example, when computer typesetting was introduced in many American newspapers, it was expected that printing operations would become more efficient. That anticipated impact eventually occurred, but other, more unanticipated impacts also happened: the number of typographical errors in the newspaper increased sharply (as the paper's personnel had to learn how to use the computer equipment properly), and the newspaper's employees often went on strike to protest the lay-off of typesetters (who were unemployed by the new technology). Here we see a general principle about the social impacts of the new communication technologies: *Impacts that are desirable, direct, and anticipated often go together, as do the undesirable, indirect, and unanticipated impacts.* The special precaution for those who introduce the new media is to see beyond the desirable, direct, and anticipated impacts, and to realize that undesirable, indirect, and unanticipated impacts usually follow. Communication research can often help detect these less-visible and less-immediate consequences of a new communication technology.

A usual investigative approach has been to gather data about the social impacts of a technological innovation by comparing a system on certain variables before, and after, the introduction of the new technology (this is a pre-post evaluation). There are many difficulties with social science research on the consequences of innovations. Our research methods are not very successful in studying a process that extends into the future. Most social research methods work best as rearview mirrors; only under special conditions can we adapt them to predict certain phenomena. One means of studying the impacts of new communication technologies is to investigate advanced systems in which one of these innovations has been used, and then extrapolate from the advanced system to other systems. In fact, this research approach is the basic design for most impacts studies of office automation. These studies show that adopting computer technology in an organization often leads to serious disruptions of the organization's work.

The impacts of computers in the office cannot be fully understood without also considering their impacts in the home. One use of home computers (in addition to playing video games) is word-processing, often in connection with work (Chapter 4). Such tele-working, with an employee working at home on a microcomputer or computer terminal for perhaps several days a week, is becoming more common. Trips to the office for conferences and other personal discussions still seem to be important socio-emotionally for most employees. Most frequently, however, an organizational member simply works at home for a few hours a night and/or on weekends. In this situation, the addictive power of computing may become a problem. The typical workaholic can now easily spend more hours of work per day by working at home. This desegregation of work and home poses difficulties for many individuals who have yet to find effective means of managing these dilemmas.

Figure 5–1 shows the numerous social impacts of the new communication technologies on work organizations. Computers can be used to supervise employees very closely. For example, in some organizations a supervisor monitors the number of keystrokes per hour made by each employee in a word-processing pool; an automatic warning message is printed on the screen of an employee whose performance is not up to an established standard. In contrast to such an electronic sweatshop, computers can be utilized as tools for employee independence and responsibility, allowing an individual to work with great autonomy. The choice between such antithetical uses of the technologies depends on how an organization decides to implement this new tool.

It is possible for an office chief to implement computer software in his or her office that will provide the chief with a "cc" of every message that is sent by the office workers to each other on an electronic messaging system. So computers *can* have a centralizing effect on communication patterns in an organization. Usually, however,

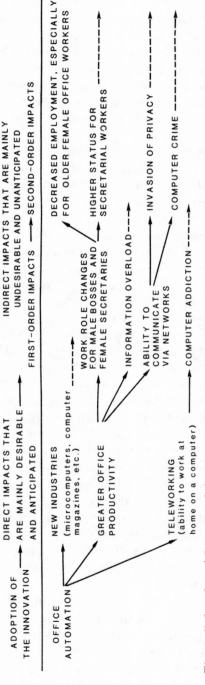

Figure 5–1. Social Impacts of the New Communication Technologies in Work Organizations

Here we have attempted to diagram certain of the more direct and more indirect social impacts of the adoption and use of office automation in work organizations in the United States. Not all of the possible future impacts are yet known because office automation was so recently introduced. With more elapsed time, more of the first-order, second-order, and third-order impacts will become clear. This diagram shows that the direct, desirable, and anticipated impacts of office automation usually are followed by the indirect, undesirable, and unanticipated impacts.

computers (especially microcomputers) have a decentralizing impact on organizational communication.

An interesting experiment in office automation occurred at Apple Computer Inc. of Cupertino, California. A few years ago, the top officials at Apple did away with secretaries and typewriters. (One of several motivations for this was to foster greater occupational and gender equality among Apple employees.) Each employee now has an Apple computer on his or her desk and another one at home; all are expected to prepare their own documents. Every fifteen to twenty employees have one "area assistant"; this person provides certain administrative-secretarial services, such as taking telephone messages, arranging meetings, and so on. Essentially, Apple has used computer technology to create a rather high degree of autonomy among its professional employees. It has also eliminated most of the usual tasks of secretaries.

One indirect consequence of office automation is changes in social status. Often managers and executives gain status by using computers, and certainly clerical staff do. Some high-level officials, however, perceive computer use as typing, which they consider a low-status job. Therefore, when a high-status individual in an organization supports the new technology, much of the resistance to office automation is likely to melt. For instance, when an electronic messaging system was introduced at Stanford University a few years ago, a photograph of the university president using the system was published in the university newspaper.

Unemployment and Social Class

One of the most serious social problems in the Information Society is unemployment. When the Industrial Society occurred, the millions of off-farm migrants that represented labor freed from agricultural production could generally be absorbed in urban-industrial jobs created in the fast-expanding cities. Today, however, there is no obvious place for former industrial workers to be absorbed. Will Information Societies be able to provide jobs in services and information for the large number of industrial workers who will be unemployed by such information technologies as robots, word-processors, and computers? The answer is: probably not.

At the heart of the transition from the Industrial Society to the Information Society is automation, which dramatically increases productivity while eliminating jobs on a massive scale. Industrial manufacturing is the chief target for computer automation, with the office close behind. One of the direct impacts of such computer-based technologies as office automation and robotics is to decrease labor, which is replaced by capital. What is to happen to the millions of office and factory workers who are being unemployed by the new technologies? Some can be put to work in the high-technology companies that produce the new communication technologies, but only a few (as high-tech industries are not very labor-intensive). During the 1980s, it is estimated that high-tech activities will generate 3 million new jobs in the U.S., but the uses of this technology will eliminate 25 million jobs. Especially hard hit will be clerical jobs; which will drop 22 percent in the insurance industry and 10 percent in banking by the year 2000 (Roessner, 1985). Further, a high proportion of high-tech work requires a high degree of formal education, which the unemployed factory worker is unlikely to possess.

In addition to the impact of new communication technologies on unemployment, they will change the social class structure of American society. A relatively small number of information workers, especially those with a high level of formal education, will form a bulge at the top of the class structure. These socioeconomic elites will be concentrated in such occupations as scientists, professors, and the owners and managers of large firms. At the other extreme of the bifurcated class structure will be very large numbers of workers in low-paying jobs. Some will have been formerly in the middle class, but have been deskilled by the new communication technologies. *Deskilling* is the process through which the new communication technologies downgrade an occupation to a lower socioeconomic status by replacing human skills with information-handling equipment.

A familiar example of deskilling is the check-out stand operators in a supermarket. Until a few years ago, such workers were typically adults who had to be proficient in memorizing hundreds of prices as well as in addition and subtraction; they used a cash register that simply recorded the results of their mental calculations. Today, the check-out clerk is likely to be a young person (often a high school student who works part-time at the supermarket). Each grocery product is identified with a bar code, which is read by a computer-based optical scanning machine; it automatically tabulates your grocery bill, so that the check-out clerk simply has to make the correct change when you pay. The benefits of this deskilling at the check-out stand have been passed along, in part, to the consumer in cheaper food prices and fewer errors in your grocery bill. But what happened to the former check-out clerks from the cash register era?

In contrast to their social impacts on unemployment and deskilling, computers and other new communication technologies also have desirable impacts, both directly and indirectly. Computers can make information work in the office more enriching, enjoyable, and productive. So the positive impacts of these technologies should not be overlooked. It's just that these overwhelmingly positive benefits come at a high human cost.

A general impact of deskilling and other social impacts of the new communication technologies is greater socioeconomic status inequality in America. The middle class is disappearing, while the upper class and the lower class are bulging. This socioeconomic distribution is in contrast to the Agricultural Society, in which the class structure was pyramid-shaped, and to the Industrial Society, in which the structure was diamond-shaped (because of a large middle class). This middle class conveyed the norms and values of society, and was a source of stability and conformity in the U.S. Now that middle class is shrinking, thanks to the social impacts of computers and other new communication technologies.

Silicon Valley Today: The Information Society of Tomorrow?*[2]

Silicon Valley is a most unusual California community in several respects: it has immense, recent riches, with an estimated 15,000 millionaires and at least two billionaires; it has a sunny, much-admired climate; and it is the world's center for producing advanced communication technologies — semiconductors, microcomputers, computer peripherals, lasers, and biotechnology products. Silicon Valley is also particularly unusual in the high degree of socioeconomic inequality that characterizes this high-technology system. At the top are the entrepreneurs, engineers, and managers who are the cream of high-tech. They live in "North County," the ten or so cities in the northern part of Santa Clara County, where an upper-middle-class life-style predominates, with large homes, late-model cars, and well-cared-for communities. Perhaps the ultimate address is Los Altos Hills, a residential community for the new-rich of Silicon Valley; the minimum lot size is two acres, and million-dollar homes are commonplace.

Until about twenty years ago, Silicon Valley was the prune capital of America, but now the plum orchards are gone, replaced by homes and high-tech parks filled with the tilt-up two-story buildings of the fast-growing microelectronics firms. The sixteen cities in Silicon Valley, once agricultural villages, have grown together to form a single metropolitan area that covers the peninsula between San Francisco and San Jose. This development was guided mainly by free-market forces, as each city and company competed with others to maximize what they wanted: jobs, low property taxes, and housing for upper-income employees. There was no centralized government to look out for the good of the entire area. Private interests outweighed public good in the rise of Silicon Valley. Understandably, not a great deal of socioeconomic equality resulted from this push-and-shove mentality of urban planning.

The underclass of Silicon Valley are predominantly Third World women who work at skilled manual jobs in microelectronics firms. They live in "South County," especially San Jose and its suburbs (for example, Alviso). Here the housing is older and often dilapidated. But these lower-cost rental units are all that the proletariat of Silicon Valley can afford. Many of the individuals who are attracted to manual work at relatively low-paid assembly-line jobs in Silicon Valley are women (men will not work for such low wages), and most of these are minorities: Spanish-speaking immigrants from Mexico, Vietnamese and other refugees from Southeast Asia, and Filipinos. Many of these Third World women are heads of single-parent households, responsible for raising children as well as working full time. There are no labor unions in Silicon Valley, and the lack of bargaining power is one reason for the relatively low pay received by these women. Their work is extremely monotonous, and this drudgery is one reason for the high job mobility, about 50 percent per year.

Not only are Third World women an important part of the work force in Silicon Valley, but the availability of such low-cost labor is also a vital element in the location of Silicon Valley microelectronics firms' plants in Malaysia, Taiwan, the Philippines, Korea, Indonesia, and other Asian nations. These off-shore locations of American microelectronics firms offer the advantage of lower wages, often only 5 to 25 percent of comparable hourly salaries for skilled manual work in California. However, the increasing automation of the semiconductor and computer assembly plants in the U.S. may eventually displace the off-shore locations.

Perhaps the Silicon Valley high-technology complex offers one scenario for the future Information Society. If so, its striking characteristic is socioeconomic inequality. Such differences stem, in part, from the prevailing emphasis upon free-market forces, which are widely applauded by high-technology entrepreneurs. Thousands of visitors travel to Silicon Valley every year in order to learn how they might emulate its success. But, to a Third World woman who does manual work in an electronics plant and who heads a single-parent household in San Jose, Silicon Valley does not look much like paradise.

Impacts on Inequality

Because the new media have strong impacts, they raise a public and scholarly concern about possible inequality in the distribution of their impacts. The framework for studying issues of equality/inequality in the effects of mass communication began with an influential article by Philip Tichenor and others (1970) in which they hypothesized that the mass media usually tend to widen differences between the information-rich and the information-poor in a mass media audience. In essence, the mass media (when they have an effect) usually have a greater effect on those individuals in the audience who are already better-informed, better-educated, and of higher socioeconomic status. Thus, information gaps are usually widened by the mass media. This tendency for information gaps to widen under most conditions is sometimes called the Matthew effect, after the biblical reference quoted on page 150.

COMMUNICATION TECHNOLOGY AND INFORMATION GAPS

The information-gaps hypothesis set off considerable research on this topic; a recent review of this literature by Celilie Gaziano (1983) included over fifty research studies, mostly dealing with the conventional mass media of radio, film, and television. Little attention has so far been directed to investigating the equality/inequality of the impacts of the new communication technologies, despite a provocative article on this topic by Natan Katzman (1974). He postulated that the new communication technologies would have the following impacts:

1. An increase in the amount of information communicated to all individuals in an audience.

2. A greater increase in the amount of information for the information-rich, than on the part of the information-poor. In other words, a new communication technology raises the level of information of all individuals, but it especially benefits the information-rich, thus widening the information-gap.

3. Information overload problems will occur, especially on the part of the information-rich, who may not be able to cope with the large volume of information they are receiving. Scientists, engineers, and managers who live and work in particularly information-

saturated environments must filter the large volume of information reaching them. Computer-based information-retrieval systems offer one means of coping with information overload problems. The information-rich are more likely to have access to such technologies for solving overload problems.

4. Newer communication technologies create new information gaps before old information gaps close. Thus, when a host of new communication technologies rapidly becomes available, as in the United States during the 1980s, the total impact is to greatly widen the information gaps in society.

Whether the new communication technologies are gap-widening or gap-narrowing is a public policy issue of great importance today. The degree of equality in the consequences of the new technologies is of central importance to policymakers who are concerned about whether a relatively expensive technology—for example, microcomputers—will widen the gap between the information-rich and the information-poor. Free market policies generally encourage wider information gaps through a hands-off position on the part of government. If the forces of free competition are allowed to operate unfettered by public policies, access to the new communication technologies will usually be unequal. The information-poor, racial/ethnic minorities, and those individuals of lower socioeconomic status will not have opportunities to use the new media that are equal to their information-rich counterparts.

In the case of such past communication technologies as television, it seemed that the new technology, in its process of diffusion, first widened the information gap in society, but eventually closed it, when everyone had adopted the innovation (Figure 5–2). The

---

Figure 5–2.   The Gap-Widening and Then Gap-Closing Nature of a New Communication Technology

Here we show the rapid rate of adoption of black-and-white television by American households during the 1950s. The information-rich adopted first, and so the immediate impact of TV diffusion was to widen the already-existing information gap. Later, when almost everyone had adopted, the impact of television was to raise the level of information for all individuals. But by this time (the 1960s), color television began to diffuse, thus creating another information gap. Today, several new communication technologies are diffusing among American households: cable TV, VCRs, and home computers. During the 1980s, they are widening the information gap in society.

SOURCES: Based in part on data from Schramm and others (1961, p. 211) and De Fleur and Ball-Rokeach (1975, p. 100).

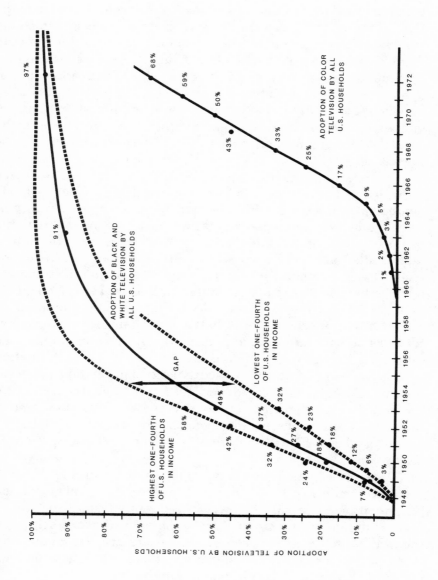

97%

91%

ADOPTION OF BLACK AND
WHITE TELEVISION BY
ALL U.S. HOUSEHOLDS

GAP

HIGHEST ONE-FOURTH
OF U.S. HOUSEHOLDS
IN INCOME

LOWEST ONE-FOURTH
OF U.S. HOUSEHOLDS
IN INCOME

58%

49%

42%

37%

32%

32%

27%

24%

23%

18%

18%

12%

7%

6%

3%

68%

59%

50%

43%

33%

25%

17%

9%

5%

3%

2%

1%

ADOPTION OF COLOR
TELEVISION BY ALL
U.S. HOUSEHOLDS

1948   1950   1952   1954   1956   1958   1960   1962   1964   1966   1968   1970   1972

100%

90%

80%

70%

60%

50%

40%

30%

20%

10%

0

ADOPTION OF TELEVISION BY U.S. HOUSEHOLDS

first-widening/then-closing sequence occurs when (and if) the technology is widely adopted; the temporary inequalities are less serious when the rate of diffusion is rapid (as in the case of television in the U.S. in the 1950s). But what about such expensive communication technology as home computers, which may never reach more than 70 to 80 percent of all American households? Further, as one new communication technology begins to reach widespread adoption, another new communication technology may be creating a new set of widened gaps (Figure 5–2). And as new communication technologies appear on the market in an increasingly rapid sequence, the overall degree of information inequality in American society may be increasing.

Gap widening will occur unless strategies are explicitly followed to prevent it because the new communication technologies—home computers, teletext and videotext systems, video cassette recorders in the home, and teleconferencing and electronic messaging in the office—are expensive. Only the socioeconomic elite can afford these new media, and they tend to adopt first (Chapter 4). These new technologies are mainly computer-based information tools that are often rather complex to utilize, and so an individual must be highly motivated to learn to use them successfully. Finally, the information-rich want specialized information, which the new communication technologies are uniquely able to provide. For instance, an individual can utilize a home computer to tap into information banks, thus obtaining such varied types of information as a congressman's daily voting record, airplane flight schedules, and up-to-the-minute stock market quotations. Perhaps an information-rich individual is in a profession like medicine and uses MEDLINE, or a lawyer and uses LEXIS—information services serving specialized occupations.

The cost of home computers dropped sharply during the mid-1980s as the industry moved from a stage of technological innovation to one of price competition. Lower-priced computers will eventually become ubiquitous in American households, and hence help close the information gaps between the more-elite and less-elite households, at least for that technology. But by then another new medium will probably open up another information gap.

The Green Thumb in Kentucky[*3]

Green Thumb was a videotext system designed especially for farmers that was provided as a joint project of the Federal Extension Service of the

USDA (U.S. Department of Agriculture), the University of Kentucky's Extension Service, and the National Weather Service. The Green Thumb decoding box contained a microprocessor and a keypad. This box was connected to a farmer's television set and telephone, enabling him to request frames of information from a local computer (located in the county extension agent's office) by placing a local telephone call and pressing a three-digit code into the keypad for each frame. Up to ten frames of information could be stored in the box, thus allowing the farmer to access Green Thumb information without tying up his telephone line for more than a minute or two. The county computer was connected to a mainframe computer at the University of Kentucky in Lexington, where most of the frames of information were input. Farm families could access about 900 frames of information on the Green Thumb system, each a colored slide dealing with farm markets, weather news, home economics information, agricultural innovations, occupational information for youth, and news of local farmers' meetings. So the Green Thumb videotext system represented a marriage of the microprocessor with telecommunications.

Two hundred of the Green Thumb boxes were distributed to as many farmers, a hundred in each of two Kentucky counties, as part of a year-long field experiment to evaluate the Green Thumb videotext system in 1981 to 1982. The project was funded by the federal government at a total cost of about $2 million. Several colleagues at Stanford University's Institute for Communication Research and I conducted an evaluation of the adoption and use of the Green Thumb system and its social impacts. We carried out personal interviews with users, nonusers, and knowledgeable informants in the two Kentucky counties, and obtained usage data that were recorded by the Green Thumb computer at the University of Kentucky.

Who Were the Green Thumb Farmers?

The 200 Kentucky farmers using the Green Thumb system operated farms averaging 625 acres in size; 63 percent of the Green Thumb farmers had at least some college. Compared to our sample of farmers not using the Green Thumb system, the users had more years of schooling and operated larger farms. These socioeconomic differences between users and non-users resulted from the selection process for the Green Thumb experiment; in each of the two Kentucky counties, local extension agents with the assistance of an advisory committee awarded the hundred Green Thumb boxes to approximately equal numbers of small, medium, and large-sized farms from among the several hundred farmers who applied (those who applied tended to be of somewhat higher socioeconomic status, on average, than farmers who did not apply). By including farmers with this range of socioeconomic status in the Green Thumb experiment, it was possible to investigate the relative usefulness of the system to larger and smaller farms, independent of the cost of the videotext system to each individual. All other videotext trials in the United States were sponsored by private firms, and charged for the information service; thus little could be learned about the usefulness of such videotext systems to individuals of lower socioeconomic status (who usually do not adopt when they must pay the rather considerable cost for a videotext system).

Among the 200 Green Thumb users, farmers who operated larger-sized farms and had adopted more agricultural innovations were heavier users of the Green Thumb system than were smaller-sized and less-innovative farmers.[4] Farmers with larger-sized operations were also less likely to stop using the Green Thumb system entirely during its year-long trial.

How Much Was the Green Thumb System Used?

The 200 farmers reported that they used the Green Thumb system an average of ninteen times per month (or about once every other day), although the data recorded by the University of Kentucky mainframe computer indicated this user-reported use was an overestimate. About 33 percent of the 200 farmers said their use decreased during the year of the Green Thumb system trial; about one-fifth of the 200 users completely stopped using Green Thumb. Clearly, the system was not an overwhelming success.

About half of our 200 respondents reported technical problems with their systems, such as a computer breakdown, or that lightning had struck their equipment, causing a power surge that "fried" the Green Thumb box in the farmer's living room. Users also complained about the lack of adequate, accurate, and up-to-date information provided by the Green Thumb system. This problem was especially characteristic of farm market information; the company inputting market data to Green Thumb interrupted their service for several months without explaining to the farmers why the interruption occurred, or even that it had occurred. The farmers just saw months-old farm prices on their Green Thumb equipment when they called up this information.

As mentioned previously, the farmers higher in socioeconomic status were higher users of the Green Thumb system. They had larger farms, more years of formal education, and were more innovative in adopting new agricultural practices. To these upscale farmers, who were more aware that they lived in an Information Society, the Green Thumb videotext system helped meet their information needs. Note that the widening of the information gap between those farmers high and low in socioeconomic status was due entirely to differential use of the new communication technology, not to *access* or to *adoption*. The original assignment of Green Thumb boxes to large, medium, and small farm (and the fact that the system was free for its first eighteen months) ensured that the usual influence of socioeconomic status on access and adoption was eliminated. Even then, smaller-sized farms used the system less than larger-sized farms, thus widening the information gap between them.

The distribution of amount of use of the Green Thumb system showed a few individuals who were very heavy users, and many individuals who were light users. This pattern of use is similar to that found for most new communication technologies, as we showed in Figure 4–3.

What Were the Effects of the Green Thumb System?

Nearly all of our respondents indicated that the main reason they applied for the Green Thumb box was to obtain a specific type of information, especially farm marketing information. Almost all of the Green Thumb users reported that marketing information was either "somewhat important" or "very important" to the operation of their farm. Futures prices, cash commodity prices, and market trends were the most-used types of marketing information. Weather information was also perceived by Green Thumb users to be very important to the operation of their farms. Radio was their major source of weather information, followed closely by television, with the Green Thumb system ranked third. In fact, the system could not provide weather information that was much more detailed or up-to-date than that available on radio or TV. But the marketing information was uniquely up-to-date.

About half of the 200 users said that Green Thumb marketing information helped them obtain higher prices for their farm products, mostly through cash marketing or grain contracting. With the Green Thumb system on their side, users of this new medium could more successfully play the grain market.

The procedure for downloading information into a farmer's Green Thumb box memory, a consequence of using telephone lines for transmission of the videotext data, discouraged use of the system for other than market and weather news. Users soon realized that each additional frame of information they ordered at log-on took seventeen seconds to download, thus lengthening their necessary wait to view the most-desired market or weather news. As a result, Green Thumb farmers usually only requested a few frames at a time in order to minimize the downloading time. Here we see an illustration of how the technical design of the Green Thumb system tended to cause barriers to its use.

Finally, the Green Thumb experiment provides understanding of why information gaps are widened by the new communication technologies. Here we see that socioeconomic status is an important determinant of the degree of use of a new medium, even when the usual effects of status on access were removed (by providing the Green Thumb system at no cost to the user). Significantly, the Green Thumb experiment was government-funded, rather than operated by a private firm (where much less attention would be given to equity issues). In the several years since the 1980–1981 Green Thumb experiment in Kentucky, a for-profit, commercial company in Cedar Rapids, Iowa, has begun to provide a Green Thumb–like video-text service to farmers. By 1985, this company had 20,000 users; each paid an initial entry fee for the service, and then paid for long-distance telephone charges. Needless to say, these are extremely large-sized, well-to-do American farmers.

Gender Inequality in Computer Use*

One of the most important inequalities being caused by the new communication technologies, especially by microcomputers, is between men and women. In fact, computers may be threatening to undo most of the gains women have been able to achieve in recent decades in the United States. In an Information Society, the much lower performance of females in math and science is a serious disadvantage for their attempts to achieve equality in socioeconomic status. This gender math gap is being worsened by computers, particularly in American high schools.

What is the historical background of gender inequality regarding computers? For years it has been known that females perform better than males in mathematics until about the sixth grade (Figure 5–3), then a marked reversal happens. The girls' performance drops below the boys', and this math gap widens with ensuing years of formal education. By the time of high school graduation, average SAT (Scholastic Aptitude Test) scores for males are fifty-nine points higher than for females, a very wide gap indeed. What happens in sixth grade that suddenly puts girls behind boys in math performance? The age of twelve or thirteen is approximately the age of puberty, when the gender expectations of parents, peers, and teachers begin to affect the behavior of adolescents. After elementary school, math and science courses are more likely to be electives, and females are less likely to enroll. Perhaps a self-fulfilling prophecy operates here; boys are popularly thought to be superior to girls in math, and, once this stereotype is accepted by boys and girls, it can affect their performance.

One of the most thorough investigations of gender inequality in computer use was conducted by Milton Chen (1985), now at Harvard University, as his doctoral dissertation at Stanford. He gathered data from a sample of 1,138 students enrolled in five high schools in the San Francisco/San Jose area. About equal numbers of boys and girls were enrolled in high school classes in English, business, history, and other classes in which microcomputers were used. But in the computer programming classes the sex ratio was about 2:1 (4.6 percent of all the female students and 8.3 percent of the males). And most of the girls dropped out of the

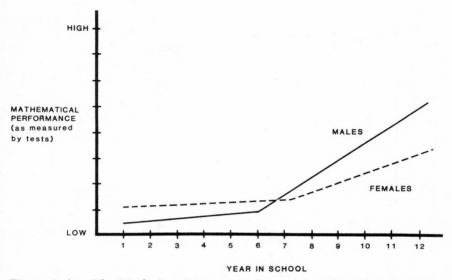

Figure 5–3. The Math Gap Between Males and Females, Which Begins at Sixth Grade, Is Widened in High School by Microcomputers

Until approximately sixth grade, girls outperform boys in mathematics, but then males push ahead, and by the end of high school, males score fifty-nine points higher than females on the SAT mathematics test. Because computer programming classes in high school are closely related to mathematics, a great deal of gender inequality occurs, thus widening the math gap. Such gender inequality has serious implications in the Information Society.

programming classes after the first semester; only 0.4 percent of girls completed three or more semesters, compared to 2.6 percent of the boys.

High school ought to be one place where gender, socioeconomic status, and ethnicity as barriers to computer access would be overcome; presumably, every student has an equal opportunity to learn how to use microcomputers. The data presented in Figure 5–4, however, show that important differences in computer experience existed on the basis of gender and ethnicity among the California high school students investigated by Chen (1985). Socioeconomic status, measured by parents' education, also contributed to wide differences in computer experience (this information gap is not shown in Figure 5–4).

Many young Americans today gain (and/or polish) their computer skills at a computer camp. Here the gender inequality is similar to that in high schools; a survey of twenty-three computer camps enrolling 5,533 students showed that females made up one-quarter of the introductory-level classes, about one-seventh of the advanced programming classes,

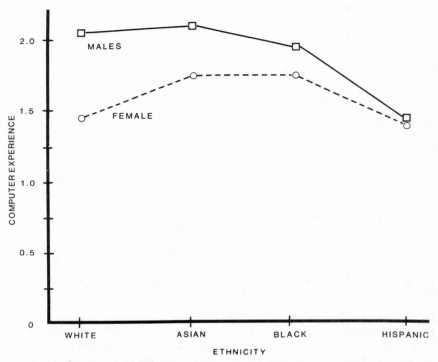

Figure 5-4. Gender Inequality in Computer Experience by Ethnic Group for California High School Students

Females have less computer experience than males for each of the four main ethnic groups in five California high schools studied by Milton Chen (1985). His measure of computer experience included use of microcomputers in elementary and junior high school, enrollment in both programming and nonprogramming courses in high school, and use of a home computer. Asian-American students of Chinese, Japanese, Korean, and Southeast Asian ancestry (who made up 9 percent of the students) achieved greater computer experience than any other ethnic group, including whites (who were 62 percent of the student body) for both males and females.

SOURCE: Based on data from Chen (1985, p. 58).

and only one-twentieth of the advanced computer courses (Hess and Miura, 1985).

Gender differences in computer experience come about for a number of reasons. Parents often encourage their sons, rather than daughters, to learn about computers. Men dominate the computer professions, and most teachers in computer courses are males. The content of most computer games features competition, violence, and fast action — themes that appeal more to boys than to girls. One of the most marked male/female differences is in aggressiveness (Maccoby and Jacklin, 1974); this means that girls are hesitant to compete with boys for access to computers, which in most American high schools are in short supply. Further, females do not really have equal access to microcomputers in high schools, especially in computer programming classes. These courses are usually taught by math teachers and require completion of mathematics courses. Computer programming courses are frequently oversubscribed, and so students are selected on the basis of their math ability. Thus, the higher math performance of males carries over into computer programming courses in high school. Finally, peer support for computer experience is stronger for boys than girls — experience and expertise are perceived more strongly by boys as a basis for respect from others (Chen, 1985).

A question might be raised as to whether high school courses in computer programming make any difference anyway. There is some evidence that little is taught of much measurable value in the typical high school programming course. For example, the 1977–1978 National Assessment of Educational Progress in the U.S. asked a national sample of 3,200 seventeen-year-olds about their computer knowledge and experience. Eleven percent of the sample reported having completed a course in computer programming. However, only half of that number were able to read a simple computer flowchart or to complete an exercise in BASIC (the computer language taught in most high school programming courses).

Nevertheless, as microcomputers become more widely adopted in American high schools and as computer teachers become better trained and more experienced, perhaps something of greater value will be taught in computer programming courses. Even at present, such courses are shaping students' attitudes toward computers and perhaps interesting them in going on to learn more about computers. So programming courses are not completely unimportant.

Can anything be done to bring about greater gender equality in computer use? California's Palo Alto High School is an illustration of what is possible. A female teacher heads the school's computer department and female engineering students from nearby Stanford University serve as teaching aides. Every effort is made to provide female high school students with appropriate role models. The result is that in this high school, equal numbers of girls and boys enroll in computer programming courses, they earn equivalent grades, and both sexes, after graduation, are equally likely to continue into computer-related study and work.

Another illustration of the potential for gender equality in computer use is provided by the state of Minnesota, which leads all U.S. states in mounting an effective school curriculum for teaching about computers. Data from a statewide sample of over 6,000 eighth-grade and eleventh-grade students in Minnesota showed no sex differences in computer literacy (that is, knowing about computers) or in test scores designed to measure ability in computer programming (Anderson and others, 1982). So evidently, when computers are ubiquitous in schools, and when there is easy access to their use, females can be equal to males in computer ability.

Computer Romance on DEAFNET*

Certain new communication technologies are of particular usefulness to communities of disabled individuals. An example are the approximately 2 million deaf and hearing-impaired in the United States. Many of the deaf cannot speak because they suffered their hearing loss before they learned to talk. Further, deaf individuals cannot hear the verbal mass media — radio, TV, and film.

In recent years, however, many television programs are captioned at the bottom of the TV screen. A deaf individual must purchase a $300 decoding box to attach to the TV set (or else buy a new television set with a built-in decoder) in order to receive the written captions, which are much like the English captions in a foreign language movie. The captions are transmitted by teletext. Today, an estimated 350,000 individuals have access to captioned television. Many of the deaf have purchased videocassette players in order to watch captioned videotapes, which they can rent or buy.

Because most of the 2 million deaf persons in the United States cannot use the telephone, computer bulletin boards are particularly useful to them. An estimated 40,000 or so deaf individuals have ancient teletype machines (many are cast-off Western Union equipment), which are interconnected with a modem and telephone lines. Many of the deaf own microcomputers, and use them to access computer bulletin boards. In recent years, the deaf community, with help from the federal government, formed DEAFNET, a national computer bulletin board. About 80 percent of the interactions on this system are within a local city, with the rest by long-distance phone lines between cities. DEAFNET provides a handy means for the hearing-impaired to talk to each other, and be heard. Here we see an application of the new communication technologies to bring about greater equality between the deaf and the nondeaf.

Betty Lou is a thirty-five-year-old lawyer in a West Coast city. She is deaf, and her speaking ability is very limited. Betty Lou has a microcomputer and a modem, and participates through DEAFNET in information exchanges with other deaf persons. In fact, she participates very heavily in the deaf network — about twenty-five to thirty hours per week. Much of her network communication is romantic in nature. For example, Betty Lou "met" Vincent, who is almost totally deaf, through the computer network. They began "talking" a year ago, originally about a routine matter, but soon they were spending several hours each evening in a very profound, frank discussion about themselves. Betty Lou says, "Even though Vincent lives in Philadelphia, we feel that our minds have met. We know each other better than couples who have lived together for years." After several months of an in-depth exchange with Vincent, Betty Lou decided to meet him in person, and traveled to Philadelphia for a two weeks' vacation. Will the two computer lovers get married? If so, they would represent a use of computer communication for socio-emotional purposes.

Information Overload

Information overload is the state of an individual or system in which excessive communication inputs cannot be processed, leading to breakdown. It is one of the main social problems resulting from the Communication Revolution. Japanese communication scholars have been most concerned with studying information overload.

In Japan, the term *"Johoka Shakai"* (or "Information Society") was coined in 1966 (Ito, 1981), and issues raised by this concept have received at least as much research attention as those in the U.S. or Western Europe. *Johoka Shakai* is defined as "a society characterized by abundant information in terms of both the stock and flow, quick and efficient distribution and transformation of information, and easy and inexpensive access to information for all members of society" (Ito, 1981). Notice that in Japan the Information Society is defined (and measured) in terms of information rather than in terms of occupational classifications or of economic contributions to the gross national product.

The research approach of Japanese communication scholars investigating the Information Society focuses on social problems (somewhat as the Chicago School of Sociology concentrated on studying the urban social problems accompanying the Industrial Society early in this century). The chief social problem investigated in Japan is information overload. In the decade from 1960 to 1970, the information supply in Japan increased by 400 percent, while the consumption of information increased only 140 percent (Bowes, 1981). The Japanese public consumed 40 percent of the information available in 1960, but only 10 percent in 1975. What caused the rapid growth in information production in Japan? Not surprisingly, more than 70 percent of the information growth was traced to such electronic communication technologies as computers, television receivers, and telephones (Bowes, 1981).

The measure of information utilized by the Japanese communication scholars is the word. They estimate the volume of words flowing through the broadcasting, publishing, and telecommunication media, and the mails. A word supplied is counted only if it is available on the premises of an individual, so that all he or she would have to do to hear or read it would be to choose to do so (such as by turning a switch or a page). Words consumed are those that an individual actually hears or reads. Surveys of the public are the source of such data about information consumption.

In the United States, certain social scientists are not much interested in the study of the Information Society concept because it has been defined in such aggregate terms that it cannot be studied with such favorite methodological tools as the survey and the experiment. In Japan, research problems (such as information overload) concerning the Information Society have been pursued at an individual or household level, as well as the societal level. For example, an information ratio has been computed as the proportion of financial expenditures for information-related activities to total household expenditures. The numerator of the ratio includes the cost of newspapers, books, and magazines, subscription fees paid for cable television, movie attendance costs, the price paid for a home computer, etc. Engel's ratio (the share of household expenditures for food) decreases with higher income, but the information ratio behaves in the opposite way: Families with higher incomes spend a higher proportion of their expenditures on information (Ito, 1981).

The information ratio can be aggregated to the national level, of course. Japanese scholars of the Information Society find that Japan's information ratio is very high for its level of personal income (Ito, 1981). However, the United States has the highest average information ratio among Western nations, and these countries have a higher information ratio than the Third World nations of Latin American, Africa, and Asia. Thus, the average information ratio of a nation seems to be highly related to how far a nation has progressed in becoming an Information Society. Indeed, the average information ratio bears a parallel to per capita income as an index of a nation's industrial development.

Following somewhat comparable procedures in the United States to those of the *Johoka Shakai* scholars in Japan, Ithiel de Sola Pool found a similar problem of information overload: "There have been extraordinary rates of growth in the transmission of electronic communications, but much lower rates of growth in the material that people actually consume" (Pool, 1983b). From 1960 to 1977, the number of words made available to Americans (over the age of ten) through seventeen public media of communication grew at the rate of 8.9 percent per year, more than double the growth rate in the gross domestic product. Words actually attended to from these media grew at just 2.9 percent, and per capita consumption of words grew only 1.2 percent per year.

The information explosion occurring in the United States is due to a much faster growth in the amount of information being supplied to the average individual than what can be absorbed. "More and more material exists, but limitations on time and energy are a controlling barrier to people's consumption of words" (Pool, 1983b). This increasing information overload means that each item of information produced faces more competition in gaining an audience. Under these conditions, certain media are growing more rapidly than others. For example, one-to-one electronic communication (that is, interactive media) are growing faster than the one-to-many mass media. "Print media are becoming increasingly expensive per word delivered while electronic media are becoming cheaper, and costs seem to predict well what is used" (Pool, 1983b).

One might question whether the relatively slower rate of consumption of information, than of information availability, is due mainly or partially to individuals' feelings of information overload. Perhaps the growing overabundance of information in society is really not a social problem; maybe citizens simply take what they need or want and leave the rest. We do not yet know very clearly which is the case. But the problem of information overload is a serious one for a society, and for the individual. This social problem is one impact of the new communication technologies. Perhaps such applications of the new communication technologies as computer-based information-retrieval systems can help us cope with information problems (Pool and Schiller, 1981).

Privacy

The issue of privacy is simply not a social problem in receiving the conventional mass media of radio, television, and film. Their one-to-many nature means that the audience member is not disclosing anything about him or herself in using these media (however, privacy issues may arise concerning private information about an individual that is inappropriately reported in the news). But once a computer is an element in a new communication system, its tremendous capacity for memory provides the possibility that privacy issues may arise. Unfortunately, to date, relatively little research attention has been given to privacy problems with the new communication technologies.

One privacy issue for such interactive cable systems as Qube concerns the monitoring of X-rated programs; the head-end computer records that households tune to such adult entertainment. What is done with this information? It could be damaging to certain individuals, such as teachers or clergymen. As was explained in Chapter 2, Warner-Amex has been very careful with the computer-recorded data about household viewing habits.

In 1983–1984, a California high school installed a computer-based telephone dialing machine that automatically phoned the parents of each student who was absent or tardy that day. The telephone delivers a short message from the school's principal, urging the parents to discuss that day's attendance problem with their child. The machine keeps dialing the parents until it reaches them, and also records their responses. Within a few months, this equipment paid for itself by decreasing daily absences by 60 percent, and thus increasing the state funds paid to the school on the basis of daily attendance. Another high school, located in an upper-middle-class community where 70 percent of the homes had adopted home computers by 1984, installed a community computer network through which parents could learn their children's daily performance in each subject.

Both of these illustrations show how computer communication can be utilized to reinforce established authority patterns. But with a different design, computer communication can be decentralizing and individual independence-creating. One might wonder if the students in these two California high schools considered the computer-based communication with their parents a violation of their privacy. At least during the first year or so of these new systems, the students did not make a strong protest.

Troublesome problems of copyright are raised by the new communication technologies. In the past, illegal copying of written material was prevented by collecting a payment for the copy at a point of gatekeeping. Take the case of a book: When you purchase a book at a bookstore, you pay an author's royalty fee as a part of the selling price (usually it is about 10 to 20 percent). But when copy machines ("Xerox" machines) became widespread in American society, how could a royalty fee be collected? Obviously, it cannot, despite the fact that the U.S. Supreme Court has upheld the illegality of such copying. When a gatekeeping point through which a certain type of information must flow is broadened by the new communication technologies, effective means of copyright protection become extremely difficult.

We see another example of this point in the case of microcomputer software programs. These information products are usually in the form of floppy disks, which are copyrighted by the software designers and sold for prices ranging from several hundred dollars (WordStar, a popular word-processing program, sold for around $400 in the early 1980s) down to $15 or $20 (for a video game). Any of these software programs can be copied by someone (a "pirate") with a microcomputer and two disk drives. In order to prevent such pirating, most software programs are encrypted with a "lock." Use of a special software program called "Locksmith" allows a pirate to break their lock, and copy the floppy disk. Here we see how the widespread use of one new technology (microcomputers) led to a threat to copyright, which in turn created a need for encryption devices, which led to technological means of overcoming such copyright protection.

Why do individuals photocopy papers out of books or journals? There is little economic reason to do so in most cases. If one includes the cost of the copier's time in operating the copy machine, the cost per copied page is about 8 or 9 cents. This figure is about twice the cost per page of buying a copy of the publication. More important as a reason for illegal copying is convenience; so much delay is involved in ordering a book at a bookstore and waiting for it to arrive, that individuals find it easier to just photocopy the book.

As we see here, one impact of the new communication technologies, especially computing, videotaping, and photocopying, is to make information more accessible. The unique and the secret becomes more difficult to keep that way. And so important new issues of individual and organizational privacy are raised by the new media.

Your Bank's Computer Knows a Lot About You*⁵

Let's say you're in Annapolis and hungry for lunch but short of money. You slip your bankcard into an automated teller and request $20. Instantly, your request is beamed to Baltimore, where a bank computer throws your query to Dayton, where another computer checks with your bank in Washington, D.C., and then tells the ATM in Annapolis to give you a $20 bill.

So, in ten seconds, thanks to the electronic banking network, you have cash to buy a crab sandwich. And a computer in Dayton knows that you skipped out of the office on a nice spring day.

Computers are easing such tasks as shopping, getting cash, and sending messages. At the same time, computer communication networks are hauling in vast amounts of personal data on the individuals who use them. Especially worrisome for the average U.S. citizen are bank computers that handle the checking of credit cards, authorize checks, and control automated teller machines. Bank officials insist they will not show an individual's personal financial data to outsiders. Your bank files can be examined by the federal government, but you must be notified that this is happening.

So the likelihood of your privacy being invaded is very unlikely. But remember that Bob Woodward and Carl Bernstein found California lawyer Donald Segretti's credit card records to be a rich source of information in pursuing the Watergate story.

Decentralization

Decentralization is the degree to which a social system is characterized by a wide sharing of power and control among the members of the system. Decisions in decentralized systems are made at the local level, such as in the lower rungs of an organization. A more centralized system is one in which key decisions are made by high officials and/or technical subject-matter experts.

To what extent do the new office communication technologies support or subvert the organizational structures as they channel communication flows? One of the anxieties expressed about the introduction of certain new technologies, such as electronic messaging, is that they will break down the constraining effect of the organization's structure on communication behavior. Will a top executive be swamped with messages when all of the organization's employees are directly connected to that official by an electronic messaging system? Will the relative ease of sending "carbons" of messages lead to problems of information overload? Does removing the constraining effect of physical distance (and the effort required to communicate across it) between two individuals in an organization greatly increase the volume of messages they exchange? Or will the new technologies, by enabling employees to work at home, increase physical barriers to face-to-face interaction among colleagues?

The general issue here is who communicates with whom, via what channels of communication, before and after the introduction of a new office communication technology. This question is made to order for communication network analysis (Rogers and Kincaid, 1981), but we know of no such investigation that has been conducted to date. The dependent variable would be the users' interpersonal network structure, as effected by the impact of a new communication technology, as measured by a pre-post research design.

There is an optional character to the new communication technologies: They offer a potential that can be used for either centralization or decentralization. Use of a new communication technology in an organization can provide individuals with more, better, and more relevant information, and thus enable them to become more autonomous decision-makers. Thereby the organization's hierarchy can be flattened. On the other hand, these same technologies can be used in a way that isolates people from each other, and that tightens the control by managers over their employees by demanding instantaneous feedback about current work progress. The way a new communication technology is used in a certain situation tells us much about an organization's climate, ideology, and problems.

The interactive communication technologies offer the possibility of decentralized communication. Everyone in a communication system can talk to everyone else. Whether or not the decentralizing nature of interactive technologies occurs depends not on the technology, but on social, political, and economic factors inherent in the context of a specific application of this technology. So again we see that the social impacts of a new communication technology often are highly dependent on the context of their use.

Pool feels that the changing technologies of communication affect the practice of free speech in a kind of "soft technological determinism" (Pool, 1983a, p. 5). For example, when communication technologies are decentralized, dispersed, and widely accessible, as in the case of microcomputers, freedom is encouraged. Technologies that are centralized, concentrated, and scarce—such as the three U.S. television networks and the early mainframe computers—discourage freedom of communication.

TELEWORKING

Optimistic predictions for the United States estimate that by 1990, 5 percent of the work force and 10 percent of the information workers will be telecommuters (Singleton, 1983, p. 196). Many of these telecommuters will work at home on a computer for two or three days a week, and travel to a central office for meetings and other duties on other workdays.

The saving of energy is a strong advantage of telecommuting. A survey of 2,000 insurance company employees in Los Angeles found they commuted an average of twenty-one miles per day, round trip; this travel totaled 12 million miles a year, and cost $2 million. Considerable time could also be saved by working at home; the 2,000 employees together spend fifty years on the L.A. freeways each calendar year (Singleton, 1983, p. 190). There are considerable benefits of telecommuting for both employees and employers (who save the cost of buildings and parking lots). Whether or not computer communication can replace transportation to work through telecommuting remains to be seen; some experts estimate that 15 million U.S. employees will work at home by 1990. At issue here is whether the organizational communication resulting from the physical decentralization of information work can be effective.

As far as the geographical decentralization of work is concerned, the new communication technologies allow independent workers to locate their activities in remote places, perhaps far away from the mother organization. Organizational coordination must then shift from a hierarchical pattern to a more decentralized mode. Such teleworking is not feasible for most office workers, as most office tasks cannot be scattered, and so they must remain concentrated in order to guarantee the unstructured interpersonal information exchange that is necessary for most organizational problem-solving and control.

Impacts of a New Medium on Older Media

A new communication technology can have social impacts at various levels: the individual, group, or organization; or at the level of an industry, a sector of the economy, or society. One important issue is how the new communication technologies impact the existing mass media. Examples are how the introduction of computers has changed newspapers, how satellite transmission has speeded up the flow of international television news, and how computer editing has altered film production. These are illustrations of impacts mainly within the existing media. In addition, a new medium can compete with, or complement, an existing medium.

One consequence of the computer on newspapers is to foster *audience segmentation*, defined as the strategy of dividing the heterogeneous audience of a mass medium into relatively more homogeneous subaudiences. Until just a few years ago, most newspaper companies delivered an identical newspaper to each of their readers. That is no longer the case, thanks to computers that can make up a localized edition of a newspaper for each suburban section in the paper's subscription area; each specialized edition, containing news of particular interest to certain readers, is not only computer-composed, but also is routed to the designated readers by computer. Thanks to computers, several big city newspapers have been able to survive the U.S. population's migration to the suburbs. With computer typesetting on their side, metropolitan dailies can beat their readers to the suburbs, thus preventing new suburban newspapers from springing up. In metropolitan Los Angeles, for example, the *Los Angeles Times*, a pioneer in using computers, bankrupted several of the suburban newspapers with the help of computer-aided segmentation (Smith, 1980, p. 150).

When newspapers also utilize satellite transmission of their text to various printing plants around the United States, they can become national in scope. Until the late-1970s, the United States did not really have a national newspaper, unlike most other countries. Today, thanks to satellites, *The New York Times*, the *Wall Street Journal*, and *USA Today* all are national dailies. For example, the *Wall Street Journal*, with a circulation of about 2 million readers, is printed at twelve plants around the U.S. Each day's issue is made up in New York, and then beamed via satellite to each printing plant, where it is modified so as to localize the content, printed, and then distributed. An Asian and a European issue of the *Wall Street Journal* are also published each day thanks to satellites and computers.

Many of the social impacts of the current crop of new communication technologies on the existing mass media cannot be fully understood for at least several more years. For this reason we go back to the case of television diffusion in the 1950s for a discussion of the social impacts of this then-new medium on such existing mass media as radio and film, and, more broadly, on many other aspects of American society.

Impacts of Television on Radio and Film in the 1950s*

The rate of adoption of black-and-white television in the United States was extremely rapid, comparable in the early years of the 1950s to the speed of diffusion of home computers, VCRs, and cable TV in the early 1980s. Wilbur Schramm and others began their discussion of how television diffused in the United States by stating: "No mass medium has ever exploded over a continent as television exploded over North America in the 1950s." (Schramm and others, 1961, p. 11). In fact, by the late 1950s, when Schramm and his colleagues were investigating the impacts of television on children in the United States, they could no longer find a community without television to use as a control group (they were forced to use a remote Canadian community).

The rapid diffusion of television in the 1950s in the U.S. had very major social impacts. Directly and immediately, leisure time use was affected, as television grabbed huge gobs of time away from radio-listening, reading, and other activities. Movie attendance went into a tailspin, the radio was pushed from the center stage of home entertainment to become just a medium for providing background music, and the authority and credibility of newspapers began to fade (Figure 5-5). Television took over the job of entertaining children from baby-sitters, movies, and playmates. Television advertising became *the* national marketing tool. Political candidates quickly found it necessary to adjust their electoral campaign strategies to utilize television spots. Consequently, the importance of political parties began to fade; television offered a more direct means of contact between candidates and their constituents. In turn, the cost of running for public office began to escalate, because of expenses of TV spots, and soon only wealthy individuals (or at least those with wealthy supporters) could afford the price. Hollywood-handsome good looks became more important in electoral success than just where a candidate stood on political issues. Thus we see how the direct effects of television echoed out through American society, causing first-generation, second-generation, and even more remote impacts.

The consequences of television on both radio and film in the United States were very powerful. Yet radio and film reacted to television in very different ways. During the 1950s, when the bottom dropped out of movie attendance in the United States, the film industry tried various strategies in order to hold on to its audience—for example such gimmicks as wider screens and three-dimensional viewing and special sound effects. None helped much; the decline in movie attendance continued. Finally, in order

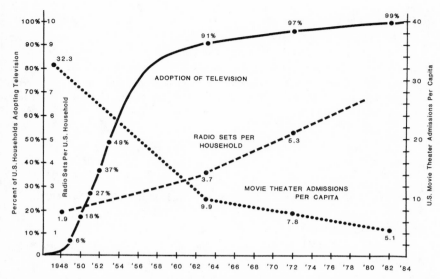

Figure 5-5. The Rapid Adoption of Television by American Households Changed the Function of Both Radio and the Movies and Sent Movie Theater Attendance into a Tailspin

These data suggest that as television adoption went up in the 1950s, movie attendance went down; it has continued to decline since about 1960, although much more slowly. Television also competed with radio, and usurped its main role in providing home entertainment. Radio changed its role to that of providing background music and news to people while they are primarily engaged in driving, working, studying, etc. Television in the United States in the 1950s provides an example of how a new medium impacts existing media. How will the new communication technologies of the 1980s affect television?

SOURCES: Based in part on data from Schramm and others (1961, p. 211) and De Fleur and Ball-Rokeach (1975, p. 100).

to attract movie goers, the moral standards governing film content were abandoned, and theater-goers were treated to frank sex portrayals. But still television continued to eat away at the film industry. Figure 5-5 shows that as the adoption of television went up, movie attendance went down, and that it continues to decline (at least on a per capita basis).

Radio, in contrast, was also affected directly by television's diffusion in the United States in the 1950s, but radio changed its function in society in order to avoid a decline. The ownership of radios by American households had gotten underway about 1925, when there were 0.14 radio sets per household. This figure rose to 0.4 in 1930, to 1.5 by 1940, and had risen to 2.3 by 1950. Thereafter, the advent of television did not slow the continued rise in this index of radio sets per household: 3.0 in 1960, 5.1 in 1970, and still rising today. But since the 1950s, radio has no longer been the dominant entertainment medium in the U.S., particularly during the evening hours. Instead, radio's role is to provide background music and news while individuals are driving, working, studying, and otherwise occupied. In short, radio caters to its audience at the times when television-viewing is inappropriate. So by changing its role in a very fundamental way, radio survived in the face of the television invasion.

Summary

The main concern of communication research over the past thirty-five years, since the field came together around the Shannon linear model of communication, has been the investigation of communication effects. *Effects* are the changes in an individual's behavior (knowledge, attitudes, or actions) that occur as the result of the transmission of a communication message.

In contrast, one of the priority concerns of communication scientists who focus on the new communication technologies is the study of *social impacts* (defined as the changes that occur to an individual or social system as a result of the adoption or rejection of an innovation). Here we categorize impacts on three dimensions:

1. *Desirable impacts* are the functional effects of an innovation on an individual or social system. *Undesirable impacts* are an innovation's dysfunctional effects on an individual or social system.

2. *Direct impacts* are the changes in an individual or social system that occur in immediate response to an innovation. *Indirect impacts* are the changes that result from the direct impacts of an innovation.

3. *Anticipated impacts* are changes caused by an innovation that are recognized and intended by the members of a social system. *Unanticipated impacts* are changes that are neither intended nor recognized.

A common research finding is that. *The social impacts of the new communication technologies that are desirable, direct, and anticipated often go together, as do the undesirable, indirect, and unanticipated impacts.*

The most important social impacts of the new communication technologies are unemployment, greater socioeconomic inequality as the information gap between the information-poor and the information-rich is widened, increased gender inequality (especially in computer use), *information overload* (defined as the state of an individual or system in which excessive communication inputs cannot be processed, leading to breakdown), concern about the invasion of privacy, *decentralization* (defined as the degree to which a social system is characterized by a wide sharing of power and control among the members of the system), and impacts on the conventional mass media such as *audience segmentation* (defined as the strategy of dividing the heterogeneous audience of a mass medium into relatively more homogeneous subaudiences).

Notes

1. *Formative evaluation* is a type of research that is conducted while an activity, process, or system is being developed or is ongoing, in order to improve its effectiveness.
2. This description is based on Everett M. Rogers and Judith K. Larsen (1984).
3. The present account draws on the Stanford University Report of the Green Thumb project (Case and others, 1981) and on Paisley (1983) and Rice and Paisley (1982).
4. Similar evidence of the widening of information gaps among farmers by a videotext system is provided by James Ettema (1984a and 1984b).
5. This case illustration draws on Peter Grier (April 19, 1984), "Automatic Tellers, Electronic Mail Raise Privacy Concerns," *Christian Science Monitor*, pp. 3–4.

CHAPTER 6

New Theory

"[As Bernard Berelson said,] on the train to progress, we all have seats facing backward. Researchers still have an opportunity to turn around and see where we may be going. If we cannot see far ahead, we may at least be able to describe what is just ahead or what we are passing by so that we can assist those who are deciding which tickets to purchase."

Ronald E. Rice, 1984

This chapter is about the theoretical implications of interactive human communication media. The early expansion of the new media in the 1980s, especially the interactive communication technologies, has brought on a "communication research revolution" in which new theory, new research methods, and new data are playing an important part. Our theme in this chapter is that this epistemological turning point demands a move away from the communication effects research that has been guided by simple linear models. Instead, more macroscopic issues should be pursued, with this broadened investigation guided by convergence models of communication. Driving the epistemological revolution in communication science is the interactivity of the new communication technologies.

Background

Around 1950, when communication research was beginning its take-off as an intellectual field, it faced a choice between two opposing paradigms. A *paradigm* is a scientific approach to some phenomena that provides model problems and solutions to a community of scholars. One theoretical alternative was represented by Claude Shannon's model of communication, which communication scholars interpreted as a linear model of one-way communication (Chapter 3). In contrast was Norbert Wiener's cybernetic model of communication as a self-correcting process through time. Wiener's more complicated model stressed the importance of feedback and implied a two-way, interactive process of information exchange. Communication scholars were well-informed about both paradigms (in fact, Wiener's 1950 book describing his cybernetic communication model was a best-seller in the U.S.).

The Shannon linear model of communication was chosen, and pursued wholeheartedly, in communication effects research. Looking backward from the viewpoint of today, one must wonder why this paradigm predominated. There are several reasons, but most important is the fact that the new communication technology in 1950 was television. During the decade of the 1950s, 91 percent of American households adopted black-and-white television. Communication scholars in large numbers turned to investigating television effects, and the Shannon model seemed to fit the study of this one-way medium best. In any event, the proper data for testing the Wiener cybernetic model were not available to communication scholars; they lacked such over-time data about feedback as a process. And they did not have the statistical or computer tools to analyze such data anyway.

Why wasn't the telephone studied by communication scholars? The reasons tell us something about the intellectual inclinations of communication scholars. The phone certainly has been around long enough (about a century) to receive research attention. In fact, that is one reason why the telephone has *not* been studied. It was already well in place (in almost every American home and place of business) prior to the rise of communication research in the 1950s. We noted (in Chapter 3) that communication research over recent decades has been attracted consistently to studying the newest communication technology. So in the 1950s, communication scientists investigated the effects of television, and ignored the ubiquitous telephone.

But such a temporal explanation is only one reason why communication scholars paid little attention to the telephone. More important, we feel, is that the telephone represents an interactive communication medium, a technology that did not fit with the primary thrust of communication research and theory toward study of the effects of one-to-many mass communication. This explanation of why communication science ignored the telephone is particularly telling today, of course, because the new communication technologies share the salient characteristic of interactivity with the phone.

In the mid-1980s, the situation for communication research is quite different from that of 1950. Now the new media represent a type of cybernetic communication. They are interactive. Over-time data about information exchange among the participants in a communication system are easily available. Understandably, interest in Wiener's cybernetic model is in a resurgence among communication scholars who are seeking to create appropriate concepts, models, and methods for investigating interactive communication.

Shifting from linear to cybernetic models will not be easy for the discipline of communication science. The change in analytical focus from the individual to the network relationship has been likened to the Renaissance paradigm shift from the earth to the sun as the center of the universe (Wilder, 1979). Just as the Copernican revolution was resisted, so will the intellectual migration to convergence models of communication.

Inadequacies of the Linear Model for Studying Interactive Communication[1]

The Latin root *communis* implies common participation or communion. Thus, the concept of communication should mean the exchange of information among the participants in a communication process. Nevertheless, as we showed in Chapter 3, communication has often been defined in quite a different way, as a one-way act through which one individual influences others. That is a linear view of communication.

The linear communication models of the 1960s—for example, David K. Berlo's S-M-C-R (source-message-channel-receiver) model (Berlo, 1960), an intellectual offspring of Shannon's theory—were useful for purposes of designing laboratory experiments that assumed one-way causality of the components of the model on communication effects. Such an assumption was justified in the study of propaganda and persuasion, especially when such messages were transmitted by the one-way mass media. These models described a simple communication *act*, but not the *process* of communication. Many important aspects of human communication do not fit linear models, and tend to be ignored by communication research based on linear models. An example is telephonic communication.

The usual approach in past effects research has been to gather data from a sample of receivers about the effects of communication on their knowledge, attitudes, and/or overt behavior. Source variables, message variables, channel variables, and/or receiver variables were manipulated by investigators as independent variables, in order to relate them to the dependent variables of communication effects. The individual receiver was usually the unit of analysis, as well as the unit of response.

CRITICISM OF THE LINEAR MODEL

Writing seventeen years after his original statement of the S-M-C-R model, Berlo (1977, p. 12) accepted the criticism of his linear model that "our view of research focusing on communication effects and our view of communication as a process have been contradictory" (Smith, 1972). Berlo (1977, p. 12) still felt that a linear model of the communication process was generally appropriate for much of human communication, although not for some of the most important communication. Berlo concluded that the intellectual interest in communication was changing, mainly from directional persuasion where a linear model may have been more satisfactory, to "communication-as-exchange."

The main problem with a linear model of communication stemmed from its basic epistemological assumptions about the nature of information, how it is transmitted, and what humans do with it. In our everyday experience there is a tendency to treat information as if it could be carried from a source to a receiver the way a bucket carries water, or a hypodermic needle injects a vaccine, or a bullet heads for a target. These analogies imply a treatment of information as if it were only a physical entity that could be moved around like other material objects. There often is a physical aspect to information: ink on paper, pixels on a computer screen, and electrical impulses in a wire. But this supposition about the nature of information considers the individual mind as an isolated entity, separate from the body, from other minds, and separate from the environment in which it exists (Bateson, 1972). The context of human communication is thus ignored in most past communication research based on a linear model and oriented to studying effects.

What is information? What makes it so important for the study of human communication? *Information* is a difference in matter-energy that affects uncertainty in a situation where a choice exists among a set of alternatives. This definition is conventional in the field of communication and is based on the work of Claude Shannon. Thus, information represents the potential for choice among a set of alternatives, each of which has some probability of occurrence greater than zero. Uncertainty is reduced when a decision is made, when one alternative is chosen from among others. So information represents the content that is exchanged among individuals when they communicate.

The assumptions that information is only a physical substance and that individual minds are contextless, led to biases in past communication theory and research:

1. To view communication as a linear one-way act (usually vertical or left-to-right), rather than a cyclical, two-way process over time (a cybernetic view)
2. To exhibit a source bias rather than to focus on the interrelationship of those who communicate
3. To focus on the objects of communication as simple, isolated physical objects, at the expense of the context in which they exist
4. To consider the primary function of communication to be one-way persuasion, rather than mutual understanding, consensus, and collective action
5. To concentrate on the psychological effects of communication on separate individuals, rather than on the social effects and the relationships among individuals within networks
6. To assume one-way mechanistic causation, rather than mutual causation, which characterizes human communication systems that are fundamentally cybernetic

These biases are interrelated and cumulative. Each supports the others in creating a coherent image of communication behavior, in spite of the limitations and problems that this image produces. When communication is perceived as one-way and persuasive, and when one takes the point of view of sources as subjects who use communication to produce a change in receivers as objects, biases toward psychological effects and mechanistic causation are created.

A Convergence Model of Communication

Communication is defined as a process in which participants create and share information with one another in order to reach a mutual understanding. A model of communication is incomplete if it only deals with a single participant's understanding of a message. Communication is always a joint occurrence, a mutual process of information sharing between two or more persons. *Communication networks* consist of interconnected individuals who are linked by patterned flows of information. Such information sharing over time leads the individuals to converge or diverge from each other in their mutual understanding of a certain topic.

Although mutual understanding is the primary function of communication, it can never be reached in any absolute sense. Several cycles of information exchange about a topic may increase mutual understanding, but will never complete it. Fortunately, for most purposes, perfect mutual understanding is not required. Generally, communication is adequate when a sufficient level of mutual understanding has been reached for the task at hand. The degree of mutual understanding can be depicted as a set of two or more overlapping circles (Venn diagrams) that represent each participant's estimate of the other's meaning, as it overlaps with the participant's meaning. The overlap, or mutual understanding, is indicated by the shaded area in Figure 6–1.

The communication process always begins with "and then . . ." to remind us that something has occurred before we begin to observe the process (Figure 6–2). Participant A may or may not consider this past before he shares information (I_1) with Participant B. B must perceive and then interpret the information that A creates to express his or her thought, and then B may respond by creating information (I_2) to share with A. A interprets this new information and then may express him or herself again with more information (I_3) about the same topic. B interprets this information, and they continue this process (I_4 . . . I_n) until one or both become satisfied that they have reached a sufficient mutual understanding of one with the other about the topic for the purpose at hand. There are no arrowheads

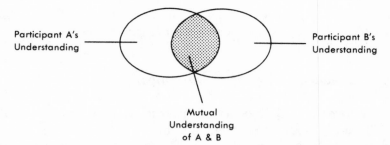

Figure 6-1. Communication as Convergence Toward Mutual Understanding

Communication is a process in which participants create and share information with one another in order to reach a mutual understanding, shown here as the overlap in understanding of two individuals.

SOURCE: Rogers and Kincaid (1981, p. 64). Used by permission.

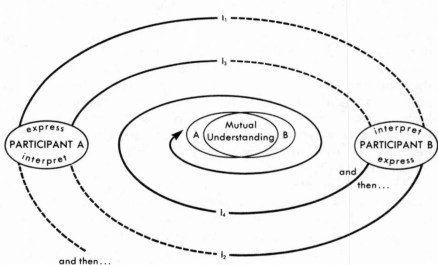

Figure 6-2. A Convergence Model of Communication

Communication, a process in which participants create and share information with one another in order to reach a mutual understanding, is a cyclical process that involves giving meaning to information that is exchanged between two or more individuals as they move toward convergence. *Convergence* is the tendency for two or more individuals to move toward one point, or for one individual to move toward another, and to unite in a common interest or focus.

SOURCE: Rogers and Kincaid (1981, p. 65). Used by permission.

"to and from" each unit of information in Figure 6–2 as the information is shared by both participants.

Convergence is always *between* two or more persons. The model compels us to study differences, similarities, and changes in human interrelationships over time. The minimal unit of analysis for communication research is the dyad, whose members are linked in an information exchange. From dyads, a communication researcher can extend his analysis to a participant's personal network[2] as well as to cliques and entire networks.

How does communication contribute to the formation of divergent factions within a network? How do such cliques affect communication? These questions are generated at the interface between the convergence model of communication and the network approach to communication research.

Units of Analysis, Variables, and Time in the Data-Cube

Three main types of data should be included in any social science analysis of human behavior: (1) the *units of analysis* (which are usually either individuals or their relationships), (2) *variables*, and (3) *time* (Figure 6–3). In most communication effects research, the first dimension on the data-cube consists of individuals, and examples of the second dimension of individual-level variables are education, attitude toward some issue, or the individual's overt behavior. Illustrative of relationship variables are who talks to whom in a system (these are network data), joint membership of two or more individuals in a group or organization, and whether a respondent has a higher or lower status than some other individual (all of these examples of relational data show that at least two or more individuals are involved). Time must be included in behavioral analysis in order to bring in process; otherwise the behavior being investigated can only be treated cross-sectionally, as if it were a "stop-action" activity.

In the typical mass communication effects survey, a large number of individuals are typically selected as respondents, and numerous variables characterizing each individual are measured. One (or more) of these dimensions are identified as dependent variables (meaning they are the behavior that a communication researcher is seeking to explain), and the other (independent) var-

UNITS OF ANALYSIS
(Individuals or relationships)   **VARIABLES**

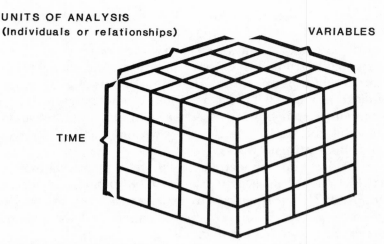

TIME

Figure 6–3. Data-Cube, Showing How a Social Science Analysis of Human Behavior Should Include: Units of Analysis (Individuals or Relationships), Variables, and Time

Raymond Cattell (1952), whose original data-cube consisted of individuals, variables, and occasions, showed that R-type factor analysis (as well as the usual uses of analysis of variance and multiple correlation) begins on the variables dimension and aggregates across individuals; Q-type factor analysis begins on the individuals' axis and aggregates across variables (and thus categorizes individuals into similar classifications). Here we utilize our version of the data-cube to argue that the unit of analysis in much communication research ought to be information-exchange relationships between people, rather than individuals (as in most past communication effects research). The new data about the new communication technologies, particularly computer-recorded data from interactive communication systems, usually have ideal qualities for research guided by the convergence model of communication: information-exchange network relationships as the units of analysis, variables that include the content of communication, and time (a type of variable that often has been missing in past research, but must be included if we are to study communication as process).

SOURCE: Rogers and Kincaid (1981, p. 80). Used by permission.

iables are correlated with them in a cross-sectional data analysis, usually utilizing a computer to handle the relatively large number of variables and units of analysis. Little or no attention is usually given to obtaining data about the respondents' relationships with other individuals, or to the over-time processual nature of the behavior being investigated. The individual is almost always the unit of analysis.

Communication Network Analysis

Communication network analysis is a method of research for identifying the communication structure in a system, in which relational data about communication flows are analyzed by using some type of interpersonal relationships as the units of analysis. This approach is particularly valuable to communication researchers because it allows them to trace specific message flows in a system, and then to compare this communication structure with the social structure of the system in order to determine how this social structure is interrelated with the communication network. The communication flow data bring life to the otherwise static nature of the social structural variables.

The essence of human behavior is the interaction through which one individual exchanges information with one or more other individuals. Any given individual in a system is likely to contact certain individuals and ignore many others (particularly when the system is large in size). As these interpersonal communication flows become patterned over time, a communication structure (or network) emerges, which is relatively stable and predictive of future human behavior. Communication network analysis describes the interpersonal linkages created by the sharing of information in the interpersonal communication structure. Previously, we defined a *communication network* as consisting of interconnected individuals who are linked by patterned communication flows.

Communication structure is the arrangement of the differentiated elements that can be recognized in the patterned communication flows in a system. One objective of communication research using network analysis is to identify this communication structure. This research objective is relatively holistic, in marked contrast to the study of communication effects on individuals. It looks for the "big picture" of communication patterns in a system.

Communication network analysis usually consists of one or more of the following research procedures:

1. To identify *cliques* (defined as subsystems whose elements interact with each other relatively more frequently than with other members of the communication system) within the total system, and then to determine how these subgroupings affect communication behavior in the system

2. To identify certain *specialized communication roles* such as *liaisons* (defined as individuals who link two or more cliques in a system, but who are not members of any clique), *bridges* (defined as individuals who link two or more cliques in a system from their position as a member of one of the cliques), and *isolates* (who are not interconnected to other individuals in the system)

3. To measure various *communication structural indexes* (communication connectedness, for example) for individuals, dyads, personal networks, cliques, or entire systems.

Early communication network scholars analyzed their data by drawing sociograms, which were usually limited to a maximum size of eighty or a hundred individuals. Each sociogram was a two-dimensional picture of the communication structure of a network. Needed were effective means of coping with the information overload caused by network data; for example, a system with 200 members can have up to 19,900 links. A sociogram of 200 individuals appears to be a hopelessly confusing tangle of arrows.

During the 1970s, several computer programs were developed for purposes of network analysis; they provide a means of identifying the communication structure of a system (that could be up to several thousand members in size) by first locating the cliques within the system. Individuals are assigned to cliques so that within-clique proximity is maximized, and between-clique proximity is minimized. *Proximity* is the degree to which two individuals' personal communication networks overlap. One or more of these computer programs for network analysis have been utilized with the same data-set, and indicate about the same communication structure (Rogers and Kincaid, 1981). So it does not seem to matter much which computer-based clique-identification techniques one uses.

A contrast between communication effects research, and communication network analysis is illustrated below.

	Communication Effects Research	*Communication Network Analysis*
1. The model upon which this research is based	Linear model	Convergence model
2. Unit of analysis	Individuals	Some type of interpersonal link

	Communication Effects Research	*Communication Network Analysis*
3. Main dependent variables	Effects of communication (knowledge, attitudes, and/or overt behavior)	(1) Who interacts with whom (2) Agreement and understanding among the individuals in the network
4. Main independent variables	Characteristics of the individual	Indices of communication structure (e.g., interconnectedness)

There are many important and useful questions about communication effects that still need to be answered. But clearly we do not need communication research in the future that is just "more of the same." Instead, the new paradigm of convergence models/network analysis should be utilized in research on the new communication technologies. And, as we show in the following chapter, the new data that can be obtained from the computer-based communication systems match almost perfectly with what is needed to test the new paradigm: message content and network links, as they occur over time.

THE INFORMATION EXCHANGE AS A UNIT OF ANALYSIS

In the study of human communication, emphasis should be placed upon information-exchange relationships, rather than on individuals, as the units of analysis. An important stumbling block in communication research is how this should be done. The intellectual paradigm consisting of the convergence model of communication and network analysis is our best guide as to how to shift to studying information-exchange relationships as the basic unit of human communication. The data for testing this paradigm are now available as computer-recorded protocols from the new interactive communication systems.

A basic question for communication theory and research is whether to study the information-exchange relationship between two or more persons as the unit of analysis, or to study the participants as individuals. The S-M-C-R model of communication, for example, explicitly separated the source and receiver. It treated messages as objects (as opposed to information) that one individual sends to another. Research based on this linear model broke up the communication process into a set of isolated variables, each associated with one of the four component elements (S, M, C, and R) in the model (plus, occasionally, feedback). The model proved useful for designing and organizing experiments on the individual effects of messages on receivers, such as the Carl Hovland–style persuasion experiments. Thousands of such persuasion studies have been completed (McGuire, 1981, p. 43).

The feedback loop, which improved the S-M-C-R model, is insufficient because it only carried (by definition) knowledge of the effects back to the source for his or her evaluation. This linear model excluded the effects of the source's messages on the source, an effect that may often be greater than the effect on the "receiver." Also excluded were the social effects of communication on other individuals who at the time were not considered the primary receivers of the source's message, even though these social effects sometimes are greater than the immediate effects on the primary receiver. Realization of such indirect effects gave rise to the two-step flow and multistep flow hypotheses in mass communication research, which deal with relaying of the source's original message to second, third, fourth, and other levels of "receivers." Communication research focused on the source's message to the initial receiver, although the recognition of multistepped flows suggested the greater complexity of the human communication process.

In comparison to the focus of past communication research on a components approach to communication effects on individuals as the units of analysis, we advocate a different approach in research on interactive communication. The convergence model of communication calls for a different approach to communication research than did the linear model of the past. First, the unit of analysis is usually the information-exchange relationship between two individuals, or some aggregation of this dyadic link to the level of the personal communication network, clique, or system.

Network data have an unusual quality. The basic data, of course, are a type of information about individual respondents: the identification of other individuals in the system with whom each respondent communicates. The solution to any network analysis problem requires that we discover certain properties of the communication structure of the system, composed of the aggregate of the individual respondents' network links. So network analysis requires an overview of an entire communication structure. Our conception of communication network analysis and of the convergence model of communication reflects an emphasis upon holistic interaction.

A basic difference in the unit of analysis is the point of departure between communication research based on a linear model and that based on a convergence model. Several other important differences follow from this shift in the unit of analysis.

We emphasize that human communication is best viewed as a holistic process. It almost always involves many other individuals than just a "source" and a "receiver," even at its most elemental level of the communication event. Past communication research has been grossly oversimplified by focusing mainly on the receiver effects of a source's message and ignoring the holistic aspects of human communication. Network analysis provides one means to study communication structure, and thus to cope more realistically with the holistic reality of human communication.

Our definition of communication as convergence implies that the sharing of information creates and defines a relationship between two or more individuals. Thus, communication behavior itself should be studied as the dependent variable in communication research. Here a main research question is "Who is linked to whom?" In comparison, most past communication research utilized communication dimensions as the independent variables to predict dependent variables that indicated such communication effects as voting, consumer behavior, aggression, and so on. These dependent variables were largely borrowed by communication scholars from other behavioral disciplines (for example, political science, marketing, psychology, etc.). And the results of communication research were often useful to these other disciplines. But until communication research began to focus on communication behavior, rather than the various effects of communication on other types of behavior, a coherent discipline of communication could not begin to emerge.

Electronic Emotion: Socio-Emotional Content in Computer Communication*

Conventional wisdom indicates that computer communication is an unlikely place to find much socio-emotional content. On the contrary, a recent investigation by Gail Love and Ronald E. Rice (1985) found that about 30 percent of the message exchanges on a computer bulletin board linking a national network of medical doctors dealt with socio-emotional content. This study is methodologically unique in that it combines a content analysis with a network analysis of the computer-recorded data.

Socio-emotional content was defined as communication messages that show solidarity, tension, tension relief, agreement, disagreement, antagonism, and giving or asking for personal information. Six weeks of the transcripts from a medical special interest group on CompuServe (a computer-based information system that individuals can access with their microcomputer and modem by telephone lines). Most of the participants in this computer bulletin board were medical doctors. This system was purged of all the accumulated messages every eight days, so Love and Rice (1985) accessed this content once each week. Their data consisted of 388 messages that were generated by 112 individuals (out of the approximately 1,000 people who had used the bulletin board during its first eighteen months of operation) over an eight-week period. The units of analysis were the 2,347 sentences in the 388 messages (content analysis is quite flexible regarding the unit of analysis, which can be a word, an utterance, a theme, etc.). Each of the sentences were coded into one of the eight types of socio-emotional content, or six types of task-related content. This content analysis was a very time-consuming task and to make it even more odious, one-tenth of the sentences were coded a second time (by a different coder) so that intercoder reliability could be computed (it was quite high, with 98 percent agreement between the two coders).

The computer bulletin board of study did not officially discourage the exchange of socio-emotional content, but its main purpose was task-related; most of the participants joined the bulletin board in order to exchange information about medical problems. Thus, it is somewhat surprising that so much socio-emotional content (30 percent of the 2,347 sentences) was found. The most frequent types of socio-emotional content were showing solidarity (18 percent of all sentences) and giving personal information (8 percent). The 70 percent of sentences classified as task-related broke down as giving nonpersonal information (57 percent), asking for nonpersonal information (6 percent), and giving an opinion (6 percent).

Love and Rice (1985) also investigated the 388 messages with the NEGOPY computer program for network analysis, treating each message as evidence of a network link. Each such computer message indicated the sender and the addressee (that is, the two individuals linked by the message exchange). Love and Rice only utilized as their data the "public" messages sent on the computer bulletin board, thus avoiding the sensitive problem of invasion of their respondents' privacy if they had also included "private" messages (on this computer bulletin board, like most others, it is possible to send a private message to an individual that other participants cannot see). The network analysis indicated there were two cliques (one of thirty-eight individuals, and the other with only three). The other seventy-one individuals were liaisons, isolates, or members of isolated dyads.

Then Love and Rice brought their two research approaches (network analysis and content analysis) together by looking at how the network communication structure was related to socio-emotional versus task-related message content. Within-clique message flows were less likely to carry socio-emotional content, while flows across clique boundaries were somewhat more likely to be socio-emotional.

Love and Rice point out that it would be possible to look at trends over time in the gradual formation of the communication structure of a computer bulletin board. When such a network begins, one would expect a large number of isolates and a lack of clique structure. Over time, thanks to the exchange of socio-emotional content among the participants in the system, more structured communication patterns would develop, as individuals form into cliques around special interests on the basis of electronic friendships. Eventually, perhaps after several years of interacting on the computer bulletin board, one would expect the cliques to stabilize.

Such conjecture raises the important question of what functions are played by socio-emotional content in human communication via computer. Perhaps socio-emotional messages build the structure of a network by constructing a communication framework or vehicle for the transmission of task-related content. If this thinking is correct, it might not be so surprising that we find a rather important share of all communication flows to be socio-emotional in nature, even on a computer bulletin board.

Time as an Essential Dimension of Communication Behavior

An adequate understanding of human communication as a process requires analysis of a series of cycles of information exchange over time. Convergence and divergence are useful concepts to describe what occurs during this process. *Convergence* is the tendency for two or more individuals to move toward one point, or for one individual to move toward another, and to unite in a common interest or focus. *Divergence* is the tendency for two or more individuals to move away or apart. In a network analysis of interactive communication over time, convergence and divergence can be utilized as dependent variables. Here we are seeking to explain a communication behavior, rather than a type of behavior (political, consumer, psychological, etc.), borrowed from some other field.

Once time has been brought into the analysis of communication behavior, we can begin to look at the sequence of changes in such behavior (an example is the stages in the formation of network patterns of communication just described). Now we can begin to see the process of communication, a type of understanding highlighted as important by definitions of communication but that has largely been ignored in past communication research. If we want to know what comes first, what else follows, etc., we must investigate communication data over time. An illustration of including the time dimension in a network analysis and content analysis of a computer bulletin board is provided by the following case illustration.

Studying Interactivity in Computer Bulletin Boards*

Past research on communication networks has almost entirely ignored the message content that flows through such networks. Such a deficiency is somewhat like building a beautiful scaffold, but then forgetting to construct a building inside. James Danowski, a professor of communication at the University of Illinois at Chicago, combined a network analysis of the participants in a computer bulletin board with a content analysis of the messages they exchanged. Danowski (1982) also took into account time, the third dimension of the data-cube (Figure 6–3). He investigated a public computer conference operating in the Boston area by the simple means of joining the network. Danowski logged-in daily and recorded all the messages on the bulletin board. The computer conference began on December 2, 1978, and Danowski selected the first ten weeks of the network's life, involving 161 message exchanges. Each message was selected, and then a search was made backward through the computer file of accumulated messages to determine if the focal message was a response to a previous message. Twenty-two pairs of message exchanges were identified in this way. Then each message was coded by two investigators in a set of twenty-five categories.

These research procedures allowed Danowski to explore such questions as: Is a change in message content associated with a change in the communication network structure? What are the stages that a computer-base communication network goes through over time? How does one message affect those that follow? His pilot study did not provide definitive answers, but indicated a useful methodology for future research.

Some flavor of the type of messages exchanged on a computer bulletin board are provided by Danowski's data:

MSG 116 IS 04 LINE(S) ON 03/06/79 FROM _____ _____
TO ALL ABOUT WANT GIRLFRIEND

DESPERATELY LONELY mathematician 33 wants compatible woman 18–35. No smoking, minimal drinking. call _____
or write to _____,
PO BOX _____.

MSG 117 IS 02 LINE(S) ON 03/07/79 FROM _____ _____
TO ALL ABOUT EXIDY SORCERER
I AM INTERESTED IN EXCHANGING EXPERIENCES, IDEAS, ETC.
WITH USERS OF THE EXIDY SORCERER.

MSG 119 IS 06 LINE(S) ON 3/09/79 _____ OF THYMESWOOD
TO _____ _____ ABOUT GIRLFRIEND

Aren't you looking a bit far from home for the love of your life?
You might have better luck if you look in California. . . . Of course,
there is the case of the two computer hackers who were married
via computer (over the ONTYME network) by this weirdo.

Note that the standard header for each message contains such data as
the sender, receiver, time, date, and topic. In short, computer-recorded interactive data are made to order for research on interactivity.

Investigating Interactivity

Interactivity was defined in Chapter 1 as the capability of new
communication systems (usually containing a computer as one component) to "talk back" to the user, almost like an individual participating in a conversation. We suggested in a previous chapter that
interactivity is a variable; some communication technologies are
relatively low in their degree of interactivity (for example, network
television), while others (such as computer bulletin boards) are more
highly interactive. But the exact degree of interactivity in a given
communication situation depends on more than just the communication technology that is involved; also important is the human
user of the technology and the context of the use. So the degree of interactivity that is present in a situation is a property of the entire
communication process.

I agree with Rafaeli, who states: "Studying interactivity is the
special intellectual niche for communication researchers" (Rafaeli,
1984). Exploring exactly what is meant in different situations by the
concept of interactivity is a high-priority task for communication
scholars. For example, Rafaeli raises the following questions:

1. Is interactivity just an antonym for unidimensionality? Or is
   interactivity something more than just two-way communication?

2. Can interactivity be present when an individual is communicating directly with a computer (rather than *via* a computer to another human)? In other words, can computers be humanlike in their responsiveness to a message?
3. Is interactivity equivalent to the interchanging of roles of sender and receiver? Or must a later message depend on information transmitted in earlier stages?
4. If interactivity is defined in terms of conversationality, how does one judge the degree to which a communication situation approaches a human conversation?

These and other theoretical issues need to be specified in order to fully explicate the concept of interactivity. Rafaeli (1984) suggests a basic proposition about interactivity for testing in future research: The greater effectiveness of highly interactive communication is associated with a higher cost of communication as measured by the amount of time required.

The basic intellectual distinction in the United States between face-to-face versus mass media communication has important disadvantages for scholars of human communication. This division of communication science comes mainly from historical causes dealing with the organization of the field; communication departments at many American universities grew out of speech departments and schools of journalism. The discipline has suffered from a lack of academic integration ever since, as it tries to overcome this trade-school orientation.

Further, much of everyday communication behavior cannot be broken down into two neat compartments on the basis of the channel used. For instance, many everyday discussions center around topics about which people are informed by the mass media. The media gather news, process it, and disseminate it via a series of interpersonal communication exchanges among reporters, editors, and technicians. If one analyzes each separate communication act, it can often be classified as involving an interpersonal channel or a mass medium. But if one analyzes the process of communication as a sequence of such unitary acts, the face-to-face versus mass media distinction becomes useless. This emphasis on communication as process is one reason why we advocated a convergence model of communication in this chapter.

This model also fits with the high degree of interactivity that characterizes the new media. When two or more individuals exchange messages via a computer-based communication system, with each message being determined in part by the previous messages in the sequence, the human participants will tend to move more closely together (or, in contrast, to diverge) in the meanings that they attach to the topic of communication. Here we see how a theoretical model is racing to catch up with the changing reality of communication technology. The rapid growth of interactive communication technologies is forcing an intellectual merger of the mass and the interpersonal approaches to communication research. It has long been overdue in the discipline of communication science.

The new media are having a powerful influence on the nature of communication research, unfreezing this field from many of its past assumptions, prior paradigms, and methods. As we have stated previously, the predominantly linear models of one-way communication effects must give way to convergence models of communication as a two-way process of information exchange, due to the interactivity of the new media. Such interactivity means that a highly individualized message content is communicated, instead of the more standardized content of print and broadcasting communication in the past. The Communication Revolution now underway in Information Societies is also a revolution in communication science, involving both models and methods.

Linear effects models of mass media effects, which have been popular in past U.S. communication research, do not function adequately in assessing the social impacts of the new media. Social impacts are much longer term and more profound in the social changes they represent than the simple, direct effects of the mass media (for instance, the effects of television on children). The new communication technologies are more powerful and more macroscopic in the nature of their impacts. For example, the social impacts of communication satellites involve a host of far-reaching changes: a boost in the rate of adoption of cable television, lower-cost long distance telephone calls, and the expansion of television reception in certain Third World nations (Chapter 8).

Only in very recent years have we begun to realize that information is becoming the fundamental resource in modern society. Thus, the study of human communication is becoming the study of society. This realization promises to put communication scholars in a key role in the new society being formed by the Communication Revolution. That is a heady potential for a scholarly field that only began to come together around 1950. Whether communication scientists are adequately prepared, theoretically and methodologically, for this challenge remains to be seen.

Summary

This chapter concerns the theoretical implications of interactive communication media. The Communication Revolution is forcing an epistemological revolution in communication research and theory. The concept of interactivity is the unique intellectual niche for communication scholars. For them to aid our further understanding of interactivity, the field of communication must change from its overwhelming focus on the study of communication effects with linear models of communication, to defining communication as convergence. In this view, *communication* is the process in which participants create and share information with one another in order to reach a mutual understanding. *Convergence* is the tendency for two or more individuals to move toward one point, or for one individual to move toward another, or to unite in a common interest or focus.

This convergence view of communication logically heads scholars toward using the tools of *communication networks*, which consist of interconnected individuals who are linked by patterned flows of information. Network analysis focuses on interpersonal relationships of information exchange as the units of analysis in communication research, rather than on individuals (as in research on communication effects). The investigation of interactive communication systems best combines network analysis (1) with content analysis of the messages that are exchanged, and (2) with time (so that communication can be studied as a process).

In 1950, the field of communication science did not have the methods *or* the data to follow the Norbert Wiener path to studying cybernetic communication. Today, thanks to computer-based communication systems, we have the data. We still don't have the fully appropriate research methods to study a communication process over time, nor a full-blown, well-developed theory, complete with concepts and middle-range propositions that could be tested with the data now available from computer-based, interactive communication systems.

Notes

1. The key ideas in the next part of this chapter are taken from Everett M. Rogers and D. Lawrence Kincaid (1981, pp. 36–78).
2. A *personal network* is the network links anchored to a single individual.

CHAPTER 7

New Research Methods

"No previous medium to videotext has been capable of monitoring its own use in such staggering detail—by user, by frames of information used, by search sequence or strategy, etc. While user surveys will still be needed to assess certain social effects of videotex use, the monitoring data may lead to new understanding of the interplay between information input and cognitive tasks like problem-solving and decision-making."

William Paisley, 1983

The new communicaton technologies affect research methods in two ways: first, they make possible the study of new research problems, which necessitate new research methods and modifications in existing methodologies; and second, the new media represent new methods for gathering data and for analyzing them. The new communication technologies are a kind of moving target for communication research in the sense that they often are changing while we study them (Johnston, 1984, p. 76). Because the new media are so young, the most appropriate designs and methods for investigating their uses and impacts are not yet very certain. So the content of the present chapter must be somewhat tentative and speculative.

New Methods for the Study of New Media

Traditional research designs are often inappropriate in ways that limit our understanding of the adoption and social impacts of the new media. Frequently, the contexts of the new media limit the degree to which the traditional research methods can be applied. The new media themselves create limitations, and opportunities, for the new kinds of research that are needed.

What are the main elements in the usual past research on communication effects? Such investigation represents the dominant approach followed by communication scholars over the past thirty-five years or so, since the Claude Shannon linear model of communication was proposed. The typical study involves an attitude change experiment or a survey of mass media effects conducted by gathering data through questionnaires or personal interviews. A large number of individual respondents (usually at least several hundred) is randomly sampled so as to provide data on a large number of variables (as many as eighty or a hundred). Quantitative analysis of this large data set is carried out using a computer. The individual is the unit of analysis and the data are analyzed cross-sectionally (they were gathered at only one point in time). Such statistical methods as correlation and analysis of variance are utilized to test the hypotheses. The entire operation has a kind of neat, clean appearance. It also represents an oversimplification of model and method, based on several questionable assumptions, as we discussed in Chapter 6. But the procedures that we have just described represent the typical study of communication effects.

THE TYPICAL EVALUATION RESEARCH DESIGN

A quite different research methodology is followed in the typical study of the new communication technologies. For example, Figure 7-1 diagrams a somewhat typical research design for studying the social impacts of a new communication technology. The main elements in the design are a sample of users of the new technology from whom data are gathered, often by means of personal interviews, both before (at t_1) and after (at t_2) the introduction of the new communication medium. This research design is a field experiment based on the kind of experimental design that behavioral scientists borrowed from classical agriculture experiments on crop yields and animal feeding.

This typical investigation of the new media is essentially an evaluation research design. In the past, evaluation research was considered a bad word by many social scientists. They looked down

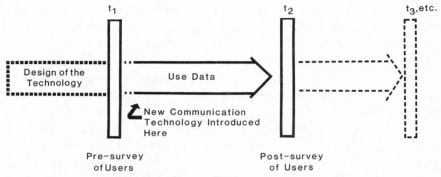

Figure 7-1. Typical Research Design for the Study of a New Communication Technology

The social impacts of a new communication technology are typically evaluated through a field experiment with a design like the above. Data are gathered on certain dependent variables (indicating social impacts) from a sample of users both before (at t_1) and after (at t_2) the introduction of a new communication medium. Communication scientists often are not involved, but they should be, in helping design the new technology prior to t_1.

SOURCE: Based somewhat on Rogers and Picot (1985).

upon research activities intended to judge the value of a system, project, or program as a kind of highly applied investigation that did not contribute much to theory development. In my opinion, the scientific worth of an evaluation study depends more on *what* is being evaluated and *how* the evaluation research is designed and conducted. Some evaluations can be designed to test a theory. For example, an evaluation of a new interactive communication system can further our understanding of the concept of interactivity, a high priority in the theoretical progress of communication science today. So to react to all evaluation research with a negative stereotype is foolish. In any event, the majority of present-day communication on the new media basically follows an evaluation research design.

If *evaluation* is defined as the use of social science research methods to gather information that can help individuals judge the value of steps that have been taken and to decide on the next steps to take (Johnston, 1984, p. 74), then questions about the adoption, use, implementation, and social impacts of the new communication technologies are often best answered by evaluation research designs. By me, there is nothing wrong about that.

Shortcomings of Existing Methods

Methodological questions, problems, and lessons accompany this type of research design for investigating the new media.

1. *Communication researchers usually enter the design and development process too late to make their maximum contribution.* Very seldom do communication researchers participate in designing a new communication technology, where engineers and technologists usually dominate.

Evaluation activities, informed by communication research, should be brought to bear early in the field experiment (prior to t_1 in Figure 7–1) in order to take full advantage of their potential contribution. Seldom is communication research explicitly involved in designing a communication technology, or in forming a strategy for adoption and implementation. The main role of communication researchers in studying the new media has been to carry out post-hoc evaluations. This passive role is very limited. Instead, communication research should be conducted much earlier in the process of designing a new technology, so that the new medium can better serve human needs and capabilities.

One illustration (there are only a few others) of communication research to design/redesign a new communication technology is Martin Elton and John Carey's year-long trial of a teletext service with forty households and in ten public sites. They concluded: "Teletext is not simple enough for many people. . . . It appears to be important for designers to understand the existing information habits and patterns of technology usage among the public. For example, we have some knowledge about how people read a magazine or change channels while watching television. Initially, it may be better to build upon these existing habits and patterns rather than try to convert consumers to an elaborate and unfamiliar information service" (Elton and Carey, 1983). This study thus led to a recommended redesign of the teletext technology, and to suggestions for diffusion/adoption strategies.

2. *Researchers who study a new communication technology are usually separate from the managers of the new media system, but they should be co-designers (or co-redesigners) of the new communication technology.* Conventional wisdom about evaluation research holds that evaluators should be independent and separate

from the system or program that they are evaluating. Politicians and other policymakers often require such separation, so as to minimize any possible protechnology bias of the evaluation. For instance, U.S. Department of Agriculture officials required that the Green Thumb system be evaluated by other than the University of Kentucky (who managed and operated the 1980–1981 Green Thumb project). So Stanford University scholars carried out an external impact evaluation, while Kentucky scholars conducted an in-house process evaluation (Bennett, 1984). The independence of evaluation researchers from the technologists who design and operate a communication technology system is thought to raise the credibility with which the research results are perceived. In many cases, such independence and separation may indeed create a healthy tension between the evaluators and the technologists, and lead to more reliable research findings about the technology's effects.

But the reality of the situation is usually that the evaluators and technologists must collaborate closely if the evaluation is to be conducted in an effective manner, and if the research findings are to be utilized to improve the communication technology through its later redesign. Often this redesign occurs during the time period (t_1 to t_2 in Figure 7–1) in which the communication technology is being evaluated; the evaluators suggest in-process needed improvements to the technologists, which may then be implemented.

It does not seem that the independence of evaluators and technologists is always ideal, even in principle. It may be crucial for both technologists and communication scientists to be involved jointly in a new communication technology; each has an important type of expertise to contribute to the new communication technology project. Perhaps it is both more realistic and advantageous for the technologists and communication scientists to be organized as a team of equals, rather than always insisting on their separate independence. The Kentucky Green Thumb program suggests that an internal evaluation has certain advantages, as does an external evaluation. Combining the two has many additional advantages.

Acceptability studies of a new communication technology face many difficulties, stemming from the basic inadequacies of available research methods to predict future behavior. For example, data from hypothetical user reactions about a new medium are of dubious value, when the respondents have not yet experienced the new communication technology. Nevertheless, acceptability studies represent one type of formative evaluation that at least involves communication scientists along with the technologists early in the process of designing a new communication technology.

3. *A control group often cannot be provided for comparison with the users of a new medium, so it is impossible to remove the effects of other variables on use of the new communication technology.* Typically, a new communication technology is introduced to an ongoing organization where the users comprise a specific department or section. Often a new office technology is launched by an innovation promoter who is an enthusiast for the new equipment. This individual is able, over a period of several months, to convince the top executives to put the new technology in place in one division of the company (often this pilot project is launched in the R & D division). A great deal of enthusiasm is created by the pilot project within the company, and soon other divisions also want to adopt. The pilot project thus functions mainly as a demonstration, rather than as an evaluation of the new communication technology. There is usually not an adequate period of time, after the pilot is started until widespread diffusion of the innovation occurs, to assess the social impacts of the new technology, especially its indirect and unanticipated consequences. Thus, to a certain extent, the organization is flying blind with the new communication technology; it doesn't really know in advance what changes will result.

The amount of use of a new computer-based communication system by an individual typically rises to an early peak in the first weeks and months of use, and then gradually declines to a somewhat lower plateau. This pattern occurs because of the initial enthusiasm of the new user, who is eager to explore the range of uses of the new medium. This novelty gradually wears off with continued use, as the new medium is incorporated into the daily routine of the user's life-style or work-style. This initially-high-but-declining pattern to use must be kept in mind when evaluating a new communication technology; a trial should be allowed ideally to run for a year or more so that use-patterns over time can be determined.

Note that there is no control group in Figure 7–1 for comparison with the users, so it is impossible to remove the effects of other variables on adoption and use of the new communication technology or to exclude certain rival hypotheses. Given the network nature of interactive communication technologies, it often is difficult to introduce new communication technologies in any other way. Perhaps another department could be selected as a control group, if it matched fairly well on certain variables. But even then, the random assignment of respondents to treatment and control groups (essential in a true experiment) is usually impossible in a study of a new medium. There are so many problems involved in having true control groups in these situations, that they are almost never utilized in evaluations of the new communication technologies. As a consequence, such evaluations tend to overestimate the effects of the new medium, because all extraneous effects that may occur are included as a disguised residual in the measured effect. Other means of control can be utilized in evaluating the new media, such as multivariate statistical control. But such an evaluation design is weaker than an experiment with control groups because all the variables to be controlled must be measured; in a well-designed experimental study, relevant variables are controlled, whether measured or not.

4. *Users of the new media often are not representative of the population of future users, so research results cannot be generalized to a wider population.* The issue of the generalizability of an evaluation experiment's results is illustrated by the respondents in an electronic mail study, who happened to be the top eighty administrators at Stanford University (Rice and Case, 1983). These administrators were selected, in part, in order to stimulate further adoption by other administrators in the university. They were not typical of the next hundred users who accepted the electronic mail system at Stanford. And how representative is Stanford University of the other organizations that were expected to adopt electronic mail in the near future? Not very.

The general lesson here is that individuals, households, and organizations that participate in a new communication technology experiment are very atypical of the larger population of potential users. Instead, the first users are usually typical of the very early adopters of an innovation: information-rich and socioeconomically advantaged (Rogers, 1983).

It is very difficult to avoid this threat (of untypical adopters) to generalizability of a new technology experiment, even when one tries. For example, in the Green Thumb project in Kentucky, the 200 users of this free (to the user) system were chosen by a local committee from the approximately 400 to 500 farmers who applied for a Green Thumb box (in response to a mailed announcement from the local county extension agent to the 2,600 farmers in the two counties of study). The local committee chose the 200 users so that they were approximately representative of three categories of farm size (small, medium, and large). This selection procedure guaranteed a range of socioeconomic status among the 200 Green Thumb users, but it introduced another bias: The small-sized farmers who volunteered to participate tended to be atypical of all small farms in the two Kentucky counties, in that they had a high degree of prior contact with their county extension agent (most small farmers do not have much extension contact). So the Green Thumb selection procedure guaranteed that small farmers were included in the study, but also tended to make these small farm–users atypical of all small farms.

There is a basic paradox in the selection procedures for participants in the trial of a new communication technology. On the one hand, if a randomly selected sample were drawn and invited to adopt the new medium, a socioeconomic bias usually occurs in who adopts versus who does not, or else in who uses the new communication technology heavily versus who uses it lightly. On the other hand, if the new medium is simply offered to all members in a system, there is a very strong socioeconomic bias in who adopts and uses. So there is no effective and practical way to overcome the higher average socioeconomic status of those individuals, households, and organizations who participate in an evaluation experiment of a new communication technology. The consequence of this bias is that the results of the trial are made to appear overly positive, and the generalizability of the results to all other members of the system is dubious. However, those socioeconomic elites who participate in the trial of a new communication medium are generally characteristic of the first adopters; to that extent the experimental results are real life.

5. *Quantitative research approaches, particularly experiments, based on variance research, seldom can provide a satisfactory understanding of the behavioral change process through which a new communication medium has impacts. Variance research* is a type of data gathering and analysis that seeks to determine the co-variances among a set of variables but not their time-order (Mohr, 1982). A t_1/t_2 pre/post design such as Figure 7–1 heads an investigation toward using (1) "difference" statistics (the t-test between two means, or analysis of variance), or (2) "correlational" statistical methods (zero-order correlation, multiple correlation and regression, or partial correlation techniques) in which the researcher seeks to determine the correlates of dependent variable(s), which usually are adoption, use, or the social impacts of the new medium. A variance approach typically assumes *linear* associations between the variables, and focuses on the social impacts of the new medium.

Variance research alone usually cannot tell us much about the time-order of the variables in a study, other than very crudely (through the t_1 to t_2 differences in a variable), and seldom can it provide a very complete understanding of the over-time process nature of the behavior impacts that are caused by a new communication technology. In this situation, a process research approach may be more appropriate.

Process research is a type of data gathering and analysis that seeks to determine the time-ordered sequence of a set of events and to explain the process by which that sequence occurs (Mohr, 1982). Data-gathering methods for process research are often more qualitative in nature (for example, participant observation, case studies, and unstructured interviewing). A special advantage of such qualitative methods is that they allow the investigator to identify unexpected variables, and to study the wider context of the user system and the new communication medium. Case studies may be particularly appropriate in the early stages of a new communication medium, as process research can be used to understand the range of uses and social impacts. For example, the 1981–1982 *Bildschirmtext* evaluation in West Germany involved qualitative data gathering via user diaries, and from an in-depth study of thirty low-income users of this videotext service (Rogers and Picot, 1985).

But process research need not necessarily be qualitative. For instance, the *Bildschirmtext* evaluation also included quantitative data gathering from a six-stage panel of 200 users, in which a core set of social impact variables was measured every four months or so. This panel design allowed tracing user behavior changes over various time periods (however, this approach is relatively obtrusive in that the repeated data gathering undoubtedly conditions the responses that are gathered). So process research can be just as quantitative as variance research; the reason it usually is not is because we lack statistical methods for analyzing over-time data that would help us understand process.

Several aspects of high-quality communication research should be included in both variance and process approaches:

1. Obtain multiple measures from several, independent sources
2. Use objective data-sources—computer-monitored data, corporate records, archival materials, etc.—rather than just individuals' self-reports as gathered in personal interviews and by questionnaires

3. Utilize unobtrusive measures, among others, so that obtaining the data does not affect the data being gathered.

Variance and process research are not necessarily mutually exclusive; a research design can include both approaches, with each providing a unique type of data. In fact, the greater the variety of research methodologies used in a research study of the new media, the better. Such triangulation always can help us learn more than any single method used alone.

New Data from New Media[2]

Computer-monitored data for a new communication system identifies the type, content, and time of each information-exchange transaction made by users of the system, which are recorded and stored automatically by the computer element in the system. Because many of the new media contain a computer as one component, these media have a memory. They usually retain a complete transcript of all messages that are exchanged through an interactive communication system. The computer-monitored data can indicate the degree of adoption/implementation, actual use of the system after the new medium has been implemented, and the social impacts of the new medium. These computer-recorded data provide a direct measure of user behavior, often complementing the data reported by users. Because the time of each message is also recorded, the ongoing, processual nature of interactive communication can be investigated with the computer-recorded data. Further, the data usually indicate who said what to whom, and so communication network links are evidenced. In short, computer-monitored data allow communication scientists to conduct the types of communication research that has long been needed, but for which we previously lacked data. The neat thing about the new data is that it fits so well with the distinctive nature of the new media: dynamic over time, specific in content, and of a network nature.

TYPES OF NEW DATA

On-line computer monitoring usually produces large data sets. A scholar usually collects a few well-defined "primitive" measures such as terminal and user identification numbers, start and end times for each message, the commands used, content of the message, response to it, errors made, etc. Then a variety of useful measures of communication behavior can be constructed from these primitive data.

Computer-monitored data can help uncover and describe patterns of use, such as the total amount of use of a new communication technology, the diversity of uses (which may be categorized as socioemotional versus task-related), and/or trends in patterns of use. For example, we found that, during the first year of the pilot project for the Green Thumb videotext system, the degree of farmer use declined sharply; this trend did not auger well for the future of the Green Thumb system.

Data on the frequency, type, and context of errors are usually provided by a new communication system, and their analysis can help identify problems in human use of a new communication technology.

Computer-monitored data also illuminate issues of timing and duration of use. For example, the elapsed time between transactions may provide useful information about an individual's "think time" associated with specific commands, and perhaps can identify certain of the more difficult-to-use features of the communication system.

Not only does the computer component in an interactive communication system provide a complete record of the content of all the information exchanges that have occurred on the system, but a computer can also be utilized to administer a questionnaire to respondents and even to administer the treatment in an experiment. For instance, in 1984, Sheizaf Rafaeli programmed a microcomputer to administer a questionnaire to respondents who were Stanford University students. In one experimental treatment, the computer was programmed to simply ask questions and record the responses, much like a paper-and-pencil questionnaire; this procedure represented the noninteractive experimental mode. In the alternative treatment condition, the microcomputer asked the same questions but in a more interactive mode; for example, a respondent could find out from the computer how everyone else had answered a question. Dr. Rafaeli thus used the computer to administer the two treatments in his experiment on the effects of the degree of interactivity in a communication system. His respondents indicated a great deal of satisfaction with their participation in the experiment; the computer made it fun for them.

COMPUTER CONTENT ANALYSIS

Content analysis is a research method that classifies message content into categories so that inferences can be drawn about the messages, their source, or their receivers. The early scholars to use content analysis, such as Harold D. Lasswell (Chapter 3), mainly utilized this method to investigate print messages (especially newspapers). After an initial burst of enthusiasm for content analysis, this method was little-used during the 1960s and 1970s, as the main attention of communication scholars turned to the study of effects. Then, in the 1980s, a resurgence of scholarly interest in utilizing content analysis occurred, with this tool being applied to the message content of computer communication, such as computer bulletin boards, electronic messaging systems, etc.

As any researcher who has carried out a content analysis will tell you, it requires a great deal of routine work. The amount of labor involved in content analyzing computer messages is especially tedious because of the de-massified nature of this communication. It is one matter to content analyze a year's issues of *The New York Times*, it is a much larger task to content analyze all the messages exchanged in a year on a computer bulletin board. Often selecting a sample of messages may reduce the information overload problem.

The use of computers as analytical tools can at times relieve some of the tedium involved in content analysis. For example, a computer can be programmed to create word-frequency lists from a text. If the communication content is not already in a computer (as is usually the case with the new media), an optical scanner can be utilized to read almost any kind of typed or printed page and transfer the text to an electronic form, such as computer tape or disk. An example of a word-frequency type of content analysis is Robert Weber's study of the Carter and Reagan platforms in the 1980 presidential election. The word *our* was the most frequently used word in both the Democratic and Republican party platforms (430 times and 347 times, respectively). *Soviet, military,* and *defense* ranked high in the Reagan platform, but not in Carter's, where the words *health, women,* and *education* were very frequent (Weber, 1985, p. 51). Of course, many other units of analysis (other than the word) can be used in content analysis: the sentence, the utterance, or a theme, for example.

ADVANTAGES OF COMPUTER-MONITORED DATA

From a logistical point of view, using the computer to collect data automatically increases the possibility of analyzing data from more respondents because such data gathering requires fewer resources than personal interviews or mailed questionnaires. Providing access to computer-collected data-bases encourages reanalysis by other researchers with differing perspectives. Replications and meta-evaluations are thus made more likely.

Unlike the information from most questionnaires and personal interviews, the collection of computer-monitored data typically involves little or no response bias from the respondents because it is unobtrusive (Webb and others, 1981). Respondents' *reports* of their communication activities may diverge widely from their *actual* communication behavior as observed or monitored (Bernard and others, 1984).

One reason why most respondents cannot provide very accurate information about communication networks is that they do not know who is linked to whom in their system, outside of themselves and perhaps some of their immediate friends. Striking evidence of this very limited knowledge of others' network links is provided by Peter Killworth and H. Russell Bernard's (1976) investigation of the communication networks among the fifty individuals on an oceanographic ship. This vessel sailed for several months at a time, a period in which those on board had no contact with the rest of the world. Under such conditions, one might expect that everyone would know who talked to whom on board the ship. Not so. The typical respondent could describe his own network links, and, with less than complete accuracy, those of his best friends (to whom he was directly linked). But beyond that, the typical respondent was a blank when asked to describe the communication networks among others on the small ship. When shown the communication structure of cliques, liaisons, and isolates resulting from a network analysis of sociometric data, the typical respondent was somewhat surprised, although these results usually made sense to him.

More generally, a series of methodological tests of respondent accuracy in providing sociometric data about communication networks lead to the conclusion: "On average, about half of what informants report is probably incorrect in some way" (Bernard and others, 1984). This indictment of self-reported data about network links is very bad news for most communication network scholars, who have rested their sophisticated network analyses almost completely on these dubious data. Fortunately, computed-recorded network data overcome these inaccuracy problems.

Computer-monitored data usually represent a complete census of information exchanges in a system, and thus allow us to investigate the communication networks among users. *Communication networks* consist of relations among interconnected individuals who are linked by patterned flows of information (Rogers and Kincaid, 1981, p. 346). Networks link individuals, organizations, and user groups with each other and with their environments. Network analysis provides one picture of a system's communication structure.

Further, a computer captures longitudinal network data, so that researchers can study the system over time. Collection of longitudinal data, particularly if it is continuous or collected at frequent discrete time intervals, allows a researcher to analyze alternative time frames. Typically, survey data has only a very few time intervals (usually just one), with each secured at a relatively high cost. Finally, a computer can administer experiments, as stated previously.

DISADVANTAGES OF COMPUTER-MONITORED DATA

The massive amounts of data collected means that a scholar must manage all those data. This information overload has serious implications for budgets, time, and expertise.

The *ethics* of using computer-monitored data about an individual may represent an invasion of privacy. In some cases, computer-monitored communication content is meant to be public, and so its analysis does not pose ethical problems. Even here, however, individuals should have the right to deny permission for access to their data.

A variety of solutions to the privacy problem are possible. As with noncomputer data gathering, respondents can be asked to sign a consent statement that explains to them how the confidentiality and anonymity of their data will be preserved. An identification number can be assigned at random to each respondent's data, so that only the investigator could match the data with the respondent. Because it is often the message content of computer communication that is confidential, rather than the network link between two or more individuals that such an information exchange represents, the content can simply be bypassed by a communication researcher. Nevertheless, the privacy problems associated with the sensitive nature of much computer communication data are one of the special difficulties entailed by the new data.

Computer-monitored data about use of a communication system have been found to be only modestly correlated with use as reported by individuals in personal interviews (Ettema, 1984b and 1985), raising troubling questions about the accuracy of such data. Of course, there is no particular reason to suspect the computer-recorded data; it is possible that respondents cannot accurately recall their exact degree of use of the system. It is also possible that the computer-recorded data may be inaccurate because of confusion or misuse of user identification numbers or for other reasons. In any event, future research is needed on the causes of such inaccuracy.

Computer-monitored data obviously do not portray the entire picture of human communication; for example, much human communication occurs on the nonverbal level. Only text messages are available for analysis. But the limited experience to data of communication researchers suggests that rich possibilities exist for investigations of interactive communication, using data obtained from computer memory.

Summary

The study of the new media by communication scientists exists in a historical, methodological, and technological context. We must understand these contexts by developing appropriate research methods for communication research on the new communication technologies. Certainly past research methods are often useful and necessary in the study of the new media, especially if appropriate modifications are made. In addition, we should design research that can use data monitored by the computer component of the interactive communication systems. These data can provide materials for process research on the new communication technologies. However, the massive amounts of data that are usually provided by computer-based communication systems cause serious problems of information overload for communication scientists.

Nevertheless, the new types of over-time, network data provided to scholars by the new communication technologies promise to take communication science in just the direction that it needs to go.

Notes

1. The early part of this chapter borrows certain ideas from Ronald E. Rice and Everett M. Rogers (1984), Everett M. Rogers and Arnold Picot (1985), and Ronald E. Rice and Christine Borgman (1982).
2. This section draws many ideas from Ronald E. Rice and Christine Borgman (1982).

CHAPTER 8

Applications of the New Communication Technologies

*"No group of social researchers could wish for more than to find their
variable at the center of transformations of work, learning, political
participation, play, and the other functions of society."*

William Paisley, 1985

This chapter describes certain of the main applications of the new
communication technologies: education and children; the home;
politics; the office; and the Third World nations of Latin America,
Africa, and Asia.

One indication of the rapid speed of the Communication Revolution is the number of computers in use. By 1985, computers were
becoming increasingly ubiquitous as they became increasingly
smaller and cheaper. In the decades since ENIAC launched the computer revolution in 1946, the number of computers went from one,
to 600 in 1956, to 30,000 in 1966, to 400,000 in 1976. By 1985, there
were over 8 million. In 1990, 50 percent of American households are
expected to have a home computer. By then, the Computer Revolution will have conquered American society. How will this revolution
change such social institutions as the family, the schools, and so
forth?

In the past, each communication technology was a separate entity—telephones, for instance, and radio and TV. Today, the distinctions among communication technologies is disappearing. Due

especially to the computer, each of these communication media are being integrated into a single communication network, which some call "compunications." Such integration occurs when a television news program is beamed by satellite to your local cable TV system, which then sends it to your home. If your system is interactive, you may respond with a signal to the head-end of the cable system, indicating your like or dislike of the news program. Your vote and others are tabulated by a computer and the results flashed on the TV screen. This communication system consists of an integrated network of satellite, cable TV, and computer technologies. Similar integration is happening in the modern office, where a computer terminal on your desk provides a word-processing function and also links you to other offices to which you can send electronic messages. These instantly transmitted messages can be printed out or stored in a computer file, thus creating a paperless office.

Communication technology extends our perceptions and knowledge and enlarges our consciousness. The computer is basic to all of the new communication technologies. It is computing capability that makes the new information technologies interactive. Interactivity in mass communication systems makes them two-way instead of one-way, as were radio and TV broadcasting and the press. Such interactivity of communication changes the nature of who controls a communication system. Each individual user of an interactive system has a large degree of control in choosing what information to request and what to avoid. Computers make possible interactive communication, and allow the individualization of information and its asynchronicity.

Education and Children[1]

In no other area of daily life is the potential of the new communication technologies having such a powerful impact as with children, both at home and at school.

Computers, especially microcomputers, provide a means to revitalize American schools. The natural affinity that children have for computers can be a powerful tool for teaching the skills needed for life in an Information Society. An exciting revolution is now underway in schools and homes in the United States to harness the teaching/learning potential of microcomputers. But this revolution still has a long way to go (as we showed in Chapter 4).

Children learn about computers with much greater ease than adults, and boys are attracted to computers more strongly than girls (Chapter 5). If children's use of computers and video games teaches them useful skills for living in an Information Society, males are getting off to a faster start. Perhaps microcomputers can provide a means of keeping girls interested in quantitative and scientific subjects, but this potential is not yet being realized, as boys presently outnumber girls three to one in learning to use computers.

Home computers in the U.S. are more accessible to socioeconomically advantaged children, thus serving to widen existing gaps between the information-poor and the information-rich. A 1982 survey of grade school children in Northern California illustrates this gap: 41 percent of the children in an upper-middle-class school reported that they had a home computer, compared to less than 1 percent in a nearby Spanish-speaking, lower-income school.

The current generation of American children will grow up with computers, as the children of the 1950s grew up with television. What is different and special about computers is that they are interactive. It is this interactive nature of computers and of communication technologies based on computers that makes using a computer so different from passively viewing television. Whether the full instructional potential of the computer will be exploited by schools and teachers, or whether computers will just go the way of instructional television, programmed learning, and language laboratories, is yet to be determined. The use of computers may affect how children think; perhaps children who grow up with computers will learn to be more logical and to think in linear sequences. We need to find out.

By 1985, it was difficult to find an American school that was not using microcomputers for teaching purposes. The typical American high school had about twenty microcomputers in 1985, with this number predicted to double each year in the immediate future. Approximately 15 percent of American households owned a microcomputer; one of the major uses was to play video games and to facilitate children's education. Over a five-year period up to 1983, the video game industry went from almost nothing to achieve $4 billion in annual sales (larger than either the film or records/music industries). During 1982, 32 billion video games were played in arcades, or 143 games for every man, woman, and child in the U.S. (Rogers and Larsen, 1984, p. 263). Most of the video game players, of course, were young boys. What is the magical attraction that computers and video games hold for children? Can this attraction, certainly demonstrated for entertainment purposes, also be utilized in educational applications? In other words, can Pac-Man fever be harnessed to teach math, English, and history? Can computers maintain their attractiveness to children over a period of time (look at what happened to the video game craze)? Does growing up with a computer change children's mental processes, such as stressing logical thinking, quantitative ability, and social alienation?

These important research questions about computers and our children have only begun to receive much attention by communication scholars. The answers are not yet understood, but demand a high priority for future inquiry.

One of the concerns of many observers who note the attractiveness of computers to today's technological child in the U.S. is that one long-term impact will be a growing unsocialness among youth. The epitome of the new computer culture is the "hacker," a computer addict who sleeps by day and works at a computer keyboard at night, feeding on junk food and the euphoria of computing. Hackers are social isolates who prefer interaction with a machine over talking to people. Hackers may be only a few hundred students at a typical university, yet the significance of hackers lies not in their present number, but in their representation of a subculture that is rapidly expanding and becoming more influential.

A new generation of hackers is being created in homes, schools, and video arcades as vast numbers of children learn to use video games and microcomputers. In only a few years, the several million school children of today who are becoming computer literates will reach college age. Those most likely to become hackers are the extremely intelligent but socially inept. Will the Computer Revolution turn large numbers of our youth into alienated hacker-nerds?

Hackers will become more of a force in the future as the U.S. becomes more of an Information Society. Today's hackers will be writing the computer programs, creating the new computer languages, and designing the new information systems to serve us in future decades. The hackers' strange ways may affect us all, as computers become more ubiquitous and the antisocial character of hackers, which seems weird today, may become more common at some future time. The hacker subculture began in the early 1970s at several high-technology universitites—MIT and Stanford to name only two—but now the phenomenon has diffused in varying degrees to many colleges, high schools, and even junior high schools.

One reason for public concern about hackers is that some computer addicts have become computer criminals. The mass media frequently carry news items about an adolescent hacker who used his or her microcomputer to illegally enter a mainframe computer network in order to transfer funds from a bank, make charge-free long-distance telephone calls, or to interface with a military defense computer. The criminal hackers seem to be motivated by a mischievous streak of anti-Establishment whimsy. Tougher state and federal laws are being enacted, along with closer FBI monitoring, to discourage computer crime, but still it grows.

It should be kept in mind, however, that the proportion of hackers who are criminals is probably quite small. Further, while the increasing spread of computers serves to encourage antisocial hacking, much computer use is actually quite social in nature. For example, children learning to use computers usually do so in groups, with one child at the keyboard and several others looking on, kibitzing, and encouraging. In fact, much of the content of computer communication among users of all ages is highly social. So the public belief that the growth of computing leads to crime and antisocial behavior is not true, except in some extreme cases.

The Home

A great deal has already been said in the earlier chapters of this book about the new communication technologies in the home: cable television, microcomputers, bulletin board networks, video cassette recorders, and video games. These new technologies are used for entertainment, information-seeking, teleshopping, and work in the American household.

The new communication technologies in the home are beginning to play an important role in banking and shopping. Most Americans have already accepted the idea of electronic banking through Visa and MasterCard—these two dominant firms between them have 120 million credit cards outstanding in the United States (many Americans carry both)—which are honored by 2 million sellers of goods and services. In Chapter 4, we described the customer resistance that banks encountered in the mid-1980s in getting more than one-third of the public to use ATMs for their banking services.

The next step in computerized banking and shopping is the "smart card," a microprocessor and memory chip embedded within a standardized credit card. A smart card is essentially a substitute for paper checks at the point of sale. The card fits into an electronic cash register that captures the data about the customer and the transaction, and then feeds it electronically through a bank to an automated clearinghouse, which credits the account of the seller and debits the account of the individual buyer. A special advantage of the smart card is that its password cannot be copied, and also that the smart card maintains tabulation of the funds remaining in the individual's account (in this sense, the smart card is somewhat like carrying your bank around in your pocket). The smart card was invented in France in 1974, and was first used in various pilot projects in 1980. However, there are presently three incompatible types of smart cards, and this lack of standardization is one factor holding up their widespread use. Smart cards are not limited to banking and shopping, of course. They could be supplied with each new auto to maintain a record in the car of all garage work performed on the car. Smart cards could also be utilized as a driver's license, as a foreign national's "green card," and, if we ever have one in the United States, as a national identification card. Hospitals and health insurance companies are already beginning to use smart cards to maintain medical records, in part to eliminate the fradulent use of health insurance.

Electronic Politics

Polling, election campaigning, and citizen participation in local and national political activities are being affected in very major ways by such technologies as computers, cable TV, satellites, and the new telephone services.

Some observers feel that the democratic process in America has been hampered by the one-way nature of the conventional mass media. Interactive communication systems permit direct voting on local, state, and national issues. The new media provide an easy means to express citizen views of government services. Interactive media can also facilitate citizen-to-citizen communication about political issues; these new media can help individuals locate others with whom they exchange information and opinions. Proposals have been made for city governments to provide public computer terminals in bus stops, shopping malls, and other public places, much like other municipal services—libraries, streets, and fire protection, for example.

Computer technology allows a political candidate to send an individualized letter to each potential voter at a modest cost. For instance, in a recent San Diego mayorial election, a specialized appeal was mailed to each citizen, depending on whether the individual was Spanish-surnamed, female, owned a Cadillac car, or was a senior citizen. A different set of political issues were stressed in that letter than in a letter sent to an individual with different characteristics. Thus the demassified nature of the new media allow complete tailoring of campaign appeals.

Much of this potential of the new media for decentralizing democracy has not yet been achieved on any widespread basis. Next steps should consist of pilot projects, with evaluations carried out by communication scientists.

The Office

The office is the primary focus of information work. Mechanization and automation reached the office relatively late, decades after these technologies were applied to agriculture, and years after they were applied to the factory. In the 1980s, it is the office where automation is starting to have a major impact.

An office secretary can usually raise typing output by about 20 percent with a word-processor because editing and retyping are so much easier. Text can be corrected without having to retype the parts of a manuscript that have not changed. The more expensive word-processors can justify the right-hand margin of the page, print in a variety of type faces, and generally make the typed page look like a thing of beauty.

By 1985, there were about a million word-processors in operation in the U.S. Yet there is more to office automation than just word-processing. Electronic mail transmits memos and letters from computerized office to computerized office, usually via telephone lines. The advantage is that messages are conveyed instantly and correspondence can be "filed" and retrieved in a computer's memory. Further, data can be put on command into graphic display on the computer screen.

Because computers are information tools, they are being utilized wherever information is input, processed, or output. Organizations are basically communication bodies (with the special characteristic of a rather high degree of formal structure, such as that imposed by authority and hierarchy), so they are natural sites for the application of a communication technology.

Considering this natural fit, it is actually surprising that office automation has not had an even greater impact to date. But although at least 20 percent of American businesses have adopted some kind of office automation, the rate of adoption has slowed in the mid-1980s. Three general factors account for this trend.

1. Although office automation promises to increase productivity, such improved performance, if it does occur, is difficult to measure. Organization leaders have to adopt these expensive technologies largely on faith. The advantages of office automation have often not been great enough to convince organizations to make the sizable initial investment they require.

2. Serious transition problems have often occurred in organizations that introduced the new information technologies. The basic requirements for proper implementation (employee training, for example), the extensive organizational changeover that the new systems demand, and the seriousness of their negative consequences (such as possible unemployment of office workers), all augur against facile adoption. In fact, implementation problems of a severe nature often typify organizations that have tried to adopt office automation.

3. Finally, computer-based office automation technologies have many shortcomings: user-unfriendliness, relatively high cost, and rapid technological change (which means that an organization must constantly purchase new machines to keep up to date).

One reason for resistance to office automation is the effect of computers on employment. If an office computer can improve a typist's efficiency by 20 percent, it means the organization can lay off 20 percent of its typists. Secretarial work represents one of the most important occupations for American women, and about 35 percent of the typical secretary's time is spent typing. Office automation threatens to add to unemployment lines. The other effect of office computers on employment is through *deskilling*, the process of job simplification by means of computer technology so that less-skilled, lower-paid workers can be substituted for more-educated, higher-paid employees (Chapter 5). Office automation can lead to deskilling if secretarial work is broken down into keyboarding on a word-processor versus other duties where shorthand dictation, accurate typing, and editing skills are required. Deskilling is an advantage to employers, but not to the skilled work force.

Not only is the office becoming automated, but so are other parts of the work organization—for example, the industrial assembly line. The word *robot* brings to mind an anthropomorphic machine that walks about on two legs, talks, and sometimes harbors sinister intentions. In reality, most present-day robots work in factories, primarily in Japan. Industrial robots bear little resemblance to humans. They look more like exotic insects, with some industrial robots resembling a human arm, with an elbow and clamplike hand. A robot possesses intelligence in the form of a computer (usually a microprocessor) programmed to control the robot in a series of repeated activities. The auto industry is presently the single most important application for industrial robots, where they work on assembly lines doing spot welding, drilling, sanding, and cutting. Robots work a twenty-four-hour shift and never go on strike.

Robots replace labor with highly capitalized information technology. Robots free humans from performing monotonous and dangerous tasks on assembly lines. Unfortunately, robots also put people out of work.

Interactive communication technologies have the potential for major impacts on work organizations, the first signs of which already exist. Office automation today seems to be having desirable, direct, and anticipated effects, but I expect that it may also lead to an expanding set of undesirable, indirect, and unanticipated consequences (as we showed in Figure 5–1).

Artificial intelligence is the ability of computers to think like humans. University scholars and R & D workers in certain high-technology firms are now devoting considerable research to the advancement of the field of artificial intelligence. Computers can consistently defeat the human chess player of average-to-expert ability. More appropriate applications of computer intelligence are to teach mathematics, diagnose medical problems, serve as skilled chemical laboratory assistants, and evaluate military tactics. In the years ahead, important breakthroughs are certain to occur in the applications of artificial intelligence, which may have even more impact on our society than the unemployment caused today by office automation and robotics.

Applications to Third World Development

Scholars consider the nations of Western Europe, North America, and Japan that represent capitalistic philosophies as the First World. The Second World consists of the nations of Eastern Europe that are under Soviet influence. The remaining nations of Latin America, Africa, and Asia are the Third World. These countries have a colonial past and did not become industrialized and wealthy during the Industrial Revolution of the past century or so. The Third World represents about 75 percent of the world's population but only about 20 percent of the world's income. The low average income in the Third World is related to the fact that half of the people are illiterate, one out of five suffers from hunger and malnutrition, and one out of four dies before the age of five.

Scholarly attention to the role of communication technology in the economic development of Third World nations was launched by two important books: Daniel Lerner's *The Passing of Traditional Society* (1958), and Wilbur Schramm's *Mass Media and National Development* (1964). The 1960s were a very upbeat era about development, and mass communication, especially the transistor radio, was expected to help solve such development problems as poverty, hunger, illness, and overpopulation. At this stage, development was defined as something that a national government did to its people. So the one-way media—radio and television—fit right into this top-down conception of development programs in health, family planning, and rural development.

But by the 1970s, the dominant paradigm of development had passed (Rogers, 1976). One reason was that development programs had not been achieving very good results in most Third World nations, even if measured by the conventional indicator of the rate of economic growth (usually computed as the annual percent of change in the per capita income). Further, while radios had spread rapidly in Latin America, Africa, and Asia, they mainly carried music and other nondevelopment programs. Television did not reach a very large audience in most Third World nations because of the relatively high cost of a TV set. Instructional television was tried in India, El Salvador, Samoa, and the Ivory Coast, but the educational impacts were discouraging.

Since the mid-1970s, the potential of development communication has become more evident. We are now more realistic about what mass communication can, and cannot, do to facilitate development. We are wiser today in assessing the possibilities for rapid development due to such new communication technologies as satellites, widespread television broadcasting, and microcomputers. These new media are causing renewed optimism about possible development in the 1980s. The past twenty-five years of development communication research in Third World settings means that we now have a body of accumulated scientific findings on which to base development communication activities.

The most important new media for Third World development are satellites, microcomputers, and VCRs. Satellites offer an especially important advantage to Third World nations: the ability to shrink the costs of communicating at a large distance. Microcomputers are too expensive for most Third World applications to date, but their potential for education and data-handling is promising. One of the surprises of the 1980s has been the rate of adoption of VCRs in certain Third World nations; a special advantage of video cassette recorders is the degree of control provided to their owners, thus often offering an alternative to government-operated television channels (Boyd and Straubhaar, 1985).

Bella Mody, a professor of international telecommunications at Michigan State University, concluded her review of experiences with communication technology in the Third World: "The different contexts of communication technology have a causal texture that is tremendously important in determining the different effects of this technology from one situation to the next" (Mody, 1985, p. 148). Certainly many mistakes have been made in recent decades when government planners transferred First World communication technologies to Third World contexts, and expected them to function similarly. They didn't.

Context is the social, psychological, political, and economic environment that surrounds a communication technology (Mody, 1985, p. 135). Media systems in the First World are mostly owned and operated by private companies in order to make a profit. The role of government in mass communication is mainly limited to a minimum of regulatory policies and to taxing. But in the Third

World, public policies are much more important; often the mass media are government-owned, they sometimes are censored, and only rarely is there an adversary relationship between the government and the press. Third World governments play a major role regarding new communication technologies; typically, they request the introduction of these new media, pay for them, and manage their implementation and use. So the transfer of a new communication technology from a First World nation—the United States—to a Third World country—say, Indonesia—means a transfer from private, for-profit ownership to government control. This ownership aspect of the context of a communication technology certainly affects how it is utilized, and with what degree of success, in a Third World setting.

One might expect that the United States would be greatly interested in assisting Third World nations with the use of new communication technologies, for reasons of enlightened altruism if not to gain a more balanced international trade and to create jobs. "The United States, with its huge internal market, advanced technology, and extensive aid programs, could be the world's leader in assisting the less developed countries in expanding and improving their telecommunications networks. Paradoxically and unfortunately, we lag far behind Western Europe and Japan in this effort" (Greenberg, 1985). Why is the U.S. not more involved in Third World communication technology, given that it is among the most profitable investments to be made in the Third World?

Mainly it is because the United States sees private business interests in selling the communication technologies in the Third World as independent from government aid policies. Other Information Societies like Japan and the Western European nations pursue public policies in their aid-giving that help them sell telecommunications in Latin America, Africa, and Asia. In 1984, U.S. international trade in communication technology went from being a positive factor to becoming a large negative factor; imports outweighed exports by several billion dollars. This crisis is leading U.S. leaders to consider policies that would assist Third World nations in purchasing satellites, telephony, and other telecommunications equipment.

Such Third World nations as Brazil, Mexico, Korea, and Singapore are pursuing a high-technology road to development, each in a somewhat different way. Essentially, these nations are trying to leap over the Industrial Society to become an Information Society. They pursue government policies that are intended to foster an indigenous microcomputer industry, a computer software industry, etc.

The key research questions here include: How can the new communication technologies be harnessed for development purposes in agriculture, health, family planning, and education? Will such hoped-for desirable impacts as more food, better health, and wider literacy outweigh such negative impacts as unemployment and the increased dependence of Third World nations on the United States, Japan, and Europe? Can the Communication Revolution enable Third World nations to leap over the Industrial Era to become Information Societies? Finally, can the distance-insensitive nature of satellite communication allow the world to be more highly interconnected in a kind of global village?

Small Media for a Big Revolution*[2]

The Iranian Revolution against the Shah in 1979 was in part a struggle against a repressive, modernizing regime by a popular, tradition-oriented movement. Not so well known is the fact that it was also a conflict between the big media versus the little media. And the little media won.

The late Shah of Iran had a love of gadgets — military equipment, nuclear energy, and communication technology. He invested heavily in broadcasting equipment, so that virtually 100 percent of the population in his large, mountainous country was reached by radio, and about 70 percent by television. Microwave relays and satellites linked the broadcasting stations. The message content carried by the broadcast media was implicitly pro-development; the TV portrayals of upper- and middle-class living conveyed a theme of modernization and an emphasis upon upward social mobility. The mass media content in Iran was uniformly pro-Shah, showing him and his family almost daily. One unintended consequence of the widespread penetration of the electronic mass media in Iran was to alienate the poor majority from the ruling elite and from the national government's drive for development.

Opposed to the Shah's regime was an informal network organized around 200,000 Islamic religious leaders (mullas) in 90,000 mosques. Although Ayatollah Khomeini and other top religious leaders were exiled abroad by the Shah, they remained in constant contact with local mullas, and hence with the people. Ayatollah Khomeini is said to have moved from living in Saudi Arabia to France when a modern, direct-dial telephone system was installed in Iran; then his followers could contact him with less fear of being apprehended by SAVAK, the Shah's dreaded secret police. From 1963 to 1979 the Ayatollah conveyed messages from exile almost daily to his followers in the form of audio tapes; these religious and political messages were widely copied by banks of tape-recording machines in the bazaars and mosques of Iran, and, after being transformed to print, were distributed by mimeographing and Xeroxing. The Ayatollah's messages were widely distributed within Iran, despite the efforts of the Shah's government to prevent the spread of this revolutionary information.

When opposition strikes against the Shah were unleashed in the Fall of 1978, dozens of newsletters and religious/political proclamations were issued daily, with Xeroxing serving as the most popular form of political communication. The huge size of the mass demonstrations (estimated at one to three million people), their extraordinary discipline, and their predominantly religious quality were evidence of the power of the small media network that flourished in spite of complete government control over the big media. As a Tehran University professor stated in October 1978: "We are struggling against autocracy, for democracy, by means of xerocracy" (Tehranian, 1979).

What is the lesson to be learned from the Iranian Revolution about the role of communication technology in the Third World? That greater recognition must be given to interpersonal channels, and to such small media as tape-recorders and Xeroxed sheets that are passed from hand to hand. It is, after all, people who can move other people to action on a mass scale. So while the Ayatollah Khomeini did not have the more powerful mass media of radio, TV, microwave relays, and communication satellites on his side, as did the Shah, the Ayatollah utilized the most appropriate communication technologies. That is an important lesson, not just for revolutionaries, but for everyone seeking to harness the power of communication technology for Third World development.

Conclusions

A brief overview of this book is provided by the following conclusions.

1. The emerging Information Societies are being driven by the new communication technologies.

2. The most distinctive single quality of the new communication technologies is their interactivity.

3. Interactive communication has certain characteristics of both face-to-face interpersonal communication and mass media communication, but it is different from either.

4. The Communication Revolution that is caused by the new

communication technologies is in turn causing a revolution in communication research, as the interactivity of the new media leads to a new paradigm for investigation.

5. Research on the new communication technologies pursues two main issues: adoption and social impacts.

6. The model for the diffusion of innovations guides research on the adoption of the new communication technologies but with some important modifications, such as the critical mass nature of early adoption.

7. The desirable, direct, and anticipated impacts of the new communication technologies often go together, as do the undesirable, indirect, and unanticipated impacts.

8. One especially important social impact of the new communication technologies is increasing inequality between the information-rich and the information-poor.

9. The exact impacts of the new communication technologies depend upon the context in which they are used.

10. The computer-recorded data available about the new communication technologies fit almost perfectly with what communication scholars need in order to study the convergence model of interactive communication: over-time process data about network information exchanges.

Notes

1. Certain of the ideas in this section are adapted from Everett M. Rogers and Judith K. Larsen (1984, pp. 252–260).
2. This case illustration is adapted from Tehranian (1979).

References

ANDERSON, R. O., and others. 1982. *Assessing Computer Literacy: Computer Awareness and Literacy*. Minneapolis: Minnesota Educational Computing Consortium.

BAKER, PAUL L. 1973. "The Life History of W. I. Thomas and Robert E. Park," *American Journal of Sociology* 79: 243–261.

BALDWIN, THOMAS F., and others. 1978. "Rockford, Ill.: Cognitive and Affective Outcomes," *Journal of Communication* 28: 168–179.

BALES, ROBERT F. 1950. *Interaction Process Analysis: A Method for the Study of Small Groups*, Cambridge, Mass.: Addison-Wesley.

BANDURA, ALBERT. 1977. *Social Learning Theory*. Englewood Cliffs, N.J.: Prentice-Hall.

BATESON, GREGORY. 1972. *Steps to an Ecology of the Mind*. New York: Ballantine.

BAUER, RAYMOND. 1964. "The Obstinate Audience: The Influence Process from the Point of View of Social Communication," *American Psychologist* 19: 319–328.

BECKER, LEE B. 1984. "Social Implications of Interactive Cable: A Decade of Research." Paper presented at the Research Forum on Advanced Wired Cities, Washington Program of the Annenberg Schools of Communications, Washington, D.C.

BELL, DANIEL. 1973. *The Coming of Post-Industrial Society: A Venture in Social Forecasting*. New York: Basic Books.

BELLO, FRANCIS. December 1953. "The Information Theory," *Fortune* 48: 136–141.

BELMAN, LARY. 1977. "John Dewey's Concept of Communication," *Journal of Communication* 27: 29–37.

BENIGER, JAMES R. 1986. *The Control Revolution: Technological and Economic Origins of the Information Society.* Cambridge, Mass.: Harvard University Press.

BENNETT, CLAUDE F. 1984. "Resolving Conflicting Demands on Evaluation: A Test Demonstration of Videotex for Farmers," *Evaluation and Program Planning* 7: 115–125.

BERELSON, BERNARD. 1959. "The State of Communication Research," *Public Opinion Quarterly* 23: 1–6.

BERLO, DAVID K. 1960. *The Process of Communication: An Introduction to Theory and Practice.* New York: Holt, Rinehart and Winston.

———. 1977. "Communication as Process: Review and Commentary," in Brent D. Ruben (ed.), *Communication Yearbook 1.* New Brunswick, N.J.: Transaction Books.

BERNARD, H. RUSSELL, and others. 1984. "The Problem of Informant Accuracy: The Validity of Retrospective Data," *American Review of Anthropology* 13: 495–517.

BOSKOFF, ALVIN. 1969. *Theory in American Sociology: Major Sources and Applications.* New York: Thomas Crowell.

BOWES, JOHN E. 1981. "Japan's Approach to an Information Society: A Critical Perspective," *Keio Communication Review* 2: 39–49.

BOYD, DOUGLAS A., and JOSEPH D. STRAUBHAAR. 1985. "Developmental Impact of the Home Video Cassette Recorder on Third World Countries," *Journal of Broadcasting and Electronic Media* 29: 5–21.

BROWNSTEIN, CHARLES N. 1978. "Interactive Cable TV and Social Services," *Journal of Communication* 28: 142–147.

BULMER, MARTIN. 1984. *The Chicago School of Sociology: Institutionalization, Diversity, and the Rise of Sociological Research.* Chicago: University of Chicago Press.

BURNS, RED, and LYNNE ELTON. 1978. "Reading, Pa.: Programming for the Future," *Journal of Communication* 28: 148–152.

CAMPBELL, JEREMY. 1982. *Grammatical Man: Information, Entropy, Language, and Life.* New York: Simon and Schuster.

CAREY, JAMES T. 1975. *Sociology and Public Affairs: The Chicago School.* Beverly Hills, Calif: Sage.

CARON, ANDRÉ H., and others. 1985. "The Process of Microcomputers in the Home: Uses and Impacts." Montreal, Université de Montreal, Department de Communication. Unpublished paper.

CASE, DONALD, and others. 1981. *Stanford Evaluation of the Green Thumb Box Experimental Videotext Project for Agricultural Extension Information Delivery in Shelby and Todd Counties, Kentucky.* Stanford Calif.: Stanford University, Institute for Communication Research. Report to the U.S. Department of Agriculture.

CATTELL, RAYMOND B. 1952. "The Three Basic Factor-Analytic Research Designs: Their Interrelations and Derivatives," *Psychological Bulletin* 49: 499–520.

CHAFFEE, STEVEN H., and JOHN L. HOCHHEIMER. 1985. "The Beginnings of Political Communication Research in the United States: Origins of the 'Limited Effects' Model," in Everett M. Rogers and Francis Balle (eds.), *The Media Revolution in America and in Western Europe*. Norwood, N.J.: Ablex.

CHAFFEE, STEVEN H., and D. G. WILSON. 1977. "Media Rich, Media Poor: Two Studies of Diversity in Agenda-Holding," *Journalism Quarterly* 54: 466–476.

CHARTERS, W. W. 1934. *Motion Pictures and Youth*. New York: Macmillan.

CHEN, MILTON. 1985. *Gender Differences in Adolescents' Uses of, and Attitudes Towards, Computers*. Ph.D. thesis, Stanford, Calif.: Stanford University.

CLARKE, PETER, and others. 1978. "Rockford, Ill.: Inservice Training for Teachers," *Journal of Communication* 28: 195–201.

COMPAINE, BENJAMIN. 1981. "Shifting Boundaries in the Information Marketplace," *Journal of Communication* 31: 132–142.

COOLEY, CHARLES HORTON. 1902. *Human Nature and the Social Order*. New York: Charles Scribner's Sons.

——. 1909. *Social Organization*. New York: Charles Scribner's Sons.

——. 1918. *The Social Process*. New York: Charles Scribner's Sons.

COSER, LEWIS A. 1984. *Refugee Scholars in America: Their Impact and Their Experience*. New Haven, Conn.: Yale University Press.

CZITROM, DANIEL J. 1982. *Media and the American Mind: From Morse to McLuhan*. Chapel Hill: University of North Carolina Press.

DAHLING, R. L. 1962. "Shannon's Information Theory: The Spread of an Idea," in Wilbur Schramm (ed.), *Studies of Innovation and of Communication to the Public*. Stanford, Calif.: Stanford University, Institute for Communication Research.

DANOWSKI, JAMES A. 1982. "Computer-Mediated Communication: A Network-Based Content Analysis Using a CBBS Conference," in Michael Burgoon (ed.), *Communication Yearbook 6*, Beverly Hills, Calif.: Sage.

DANOWSKI, JAMES A., and PAUL EDISON SWIFT. 1985. "Crisis Effects on Intraorganizational Computer-Based Communication," *Communication Research* 12: 251–270.

DE FLEUR, MELVIN L., and SANDRA BALL-ROKEACH. 1975. *Theories of Mass Communication*, 3rd ed. New York: David McKay.

DENZIN, NORMAN K. 1984. "On Interpreting an Interpretation," *American Journal of Sociology* 89: 1426–1432.

DEUTSCH, KARL W. 1963. *The Nerves of Government: Models of Political Communication and Control.* New York: Free Press.

DOMINICK, JOSEPH R. 1983. *The Dynamics of Mass Communication.* Reading, Mass.: Addison-Wesley.

DUPUY, JEAN-PIERRE. 1980. "Myths of the Information Society," in Kathleen Woodward (ed.), *The Myths of Information: Technology and Postindustrial Culture.* Madison, Wis.: Coda Press.

DUTTON, WILLIAM, and others. 1985. "Computing in the Home: A Research Paradigm," *Computers and the Social Sciences* 1: 5–18.

DUTTON, WILLIAM, and others. 1986. "The Diffusion and Social Impacts of Computing Among Households: A Meta-Analysis of Survey Research." Paper presented at the International Communication Association, Chicago.

DUTTON, WILLIAM, and others (eds.). In press. *Shaping the Future of Communications: National Visions and Wired Cities Ventures.*

ELTON, MARTIN, and JOHN CAREY. 1983. "Computerizing Information: Consumer Reaction to Teletext," *Journal of Communication* 33: 162–173.

ELTON, MARTIN C. J. and others (eds.). 1978. *Evaluating New Telecommunication Services.* New York: Plenum Press.

ETTEMA, JAMES S. 1984a. "Three Phases in Creation of Information Inequalities: An Empirical Assessment of a Prototype Videotex System," *Journal of Broadcasting* 28: 383–395.

———. 1984b. "Videotex for Market Information: A Survey of Prototype Users," in Jerome Johnston (ed.), *Evaluating the New Information Technologies.* San Francisco: Jossey-Bass.

———. 1985. "Explaining Information System Use with System-Monitored vs. Self-Reported Use Measures," *Public Opinion Quarterly* 49: 381–387.

FARIS, ROBERT E. L. 1970. *Chicago Sociology, 1920–1932.* Chicago: University of Chicago Press.

FESTINGER, LEON (ed.). 1980. *Retrospective on Social Psychology.* New York: Oxford University Press.

FINN, SETH, and DONALD F. ROBERTS. 1984. "Source, Destination, and Entropy: Reassessing the Role of Information Theory in Communication Research," *Communication Research* 11: 453–476.

FISHER, RONALD A. 1958. *Statistical Methods for Research Workers*, 13th ed. New York: Haefner.

FRAZIER, P. JEAN, and CECILIE GAZIANO. 1979. *Robert E. Park's Theory of*

News, Public Opinion and Social Control. Lexington, Ky.: Journalism Monographs.

FREDIN, ERIC S. 1983. "The Context of Communication: Interactive Telecommunication, Interpersonal Communication, and Their Effect on Ideas," *Communication Research* 10: 553–581.

GALLUP, GEORGE. 1930. "A Scientific Method for Determining Reader Interest," *Journalism Quarterly* 7: 1–13.

GAZIANO, CECILIE. 1983. "The Knowledge Gap: An Analytical Review of Media Effects," *Communication Research* 10: 447–486.

GEORGOUDI, MARIANTHI, and RALPH L. ROSNOW. 1985. "The Emergence of Contextualism," *Journal of Communication*, 35: 76–88.

GLASS, GENE V., and others. 1975. *Decision and Analysis of Time-Series Experiments*. Boulder, Colo.: Associated University Press.

GREENBERG, ALLEN. 1985. "Impasse?: The U.S. Stake in Third World Telecommunications Development," *Journal of Communication* 35: 42–49.

GRIER, PETER. April 19, 1984. "Automatic Tellers, Electronic Mail Raise Privacy Concerns," *Christian Science Monitor*, pp. 3–4.

HARTLEY, R. V. L. 1928. "Transmission of Information," *Bell System Technical Journal* 17: 535–550.

HEIMS, STEVE P. 1977. "Gregory Bateson and the Mathematicians: From Interdisciplinary Interaction to Societal Functions," *Journal of the History of the Behavioral Sciences* 13: 141–159.

———. 1980. *John von Neumann and Norbert Wiener: From Mathematics to the Technologies of Life and Death*. Cambridge, Mass.: MIT Press.

HERTZBERG, DANIEL. February 21, 1985. "If Carrots Don't Persuade People to Use ATM's, Banks Go for Sticks," *Wall Street Journal*.

HESS, ROBERT, and I. T. MIURA. 1985. "Gender Differences in Enrollment in Computer Camps and Classes: The Extracurricular Acquisition of Computer Training," *Sex Roles: A Journal of Research* 13.

HIMMELWEIT, HILDA A., and others. 1958. *Television and the Child*. London: Oxford University Press.

HINKLE, ROSCOE, and GISELA HINKLE. 1954. *The Development of Modern Sociology*. New York: Random House.

HOVLAND, CARL I., and others. 1949. *Experiments on Mass Communication*. Princeton, N.J.: Princeton University Press.

HOVLAND, CARL I., and others. 1953. *Communication and Persuasion*. New Haven, Conn.: Yale University Press.

HYMAN, HERBERT H., and PAUL B. SHEATSLEY, 1947. "Some Reasons Why Information Campaigns Fail," *Public Opinion Quarterly* 11: 412–423.

Ito, Youichi. 1981. "The 'Johoka Shakai' Approach to the Study of Communication in Japan," *Keio Communication Review* 1: 13–40.

Jay, Martin. 1984. *Adorno*. Cambridge, Mass.: Harvard University Press.

Johansen, Robert. 1984. "Foreword," in Ronald E. Rice and Associates (eds.), *The New Media: Communication, Research, and Technology*. Beverly Hills, Calif.: Sage.

Johansen, Robert and others. 1979. *Electronic Meetings: Technical Alternatives and Social Choice*. Reading, Mass.: Addison-Wesley.

Johnston, Jerome, (ed.). 1984. *Evaluating the New Information Technologies*. San Francisco: Jossey-Bass.

Katz, Elihu. 1983. "The Return of the Humanities and Sociology," *Journal of Communication* 33: 51–52.

Katz, Elihu, and Paul F. Lazarsfeld, 1955. *Personal Influence: The Part Played by People in the Flow of Mass Communications*. New York: Free Press.

Katzman, Natan. 1974. "The Impact of Communication Technology: Some Theoretical Premises and Their Implications," *Ekistics* 225: 125–130.

Kay, Peg. 1978. "Policy Issues in Interactive Cable Television," *Journal of Communication* 28: 202–208.

Killworth, Peter, and H. Russell Bernard. 1976. "Informant Accuracy in Social Network Data," *Human Organization* 35: 269–286.

Klapper, Joseph T. 1960. *The Effects of Mass Communication*. New York: Free Press.

Kuhn, Thomas S. 1970. *The Structure of Scientific Revolutions*. Chicago: University of Chicago Press.

Kuklick, Henrika. 1984. "The Ecology of Sociology," *American Journal of Sociology* 85: 1433–1440.

Kwiatek, Kathy Krendl. 1982. "New Ideas in the Workplace: Learning from Interactive Television," *Journal of Educational Technology Systems*, pp. 117–129.

Lasswell, Harold D. 1927. *Propaganda Technique in the World War*. New York: Knopf.

———. 1953. *World Politics and Personal Insecurity*. New York: Whittlesey House.

———. 1954. "The Structure and Function of Communication in Society," in Lyman Bryson (ed.), *The Communication of Ideas*. New York: Cooper Square.

Lazarsfeld, Paul F., and others. 1944. *The People's Choice*. New York: Duell, Sloan, and Pearce.

LAZARSFELD, PAUL F., and FRANK N. STANTON (eds.). 1941. *Radio Research, 1941*. New York: Duell, Sloan, and Pearce.

—— (eds.). 1944. *Radio Research, 1941–1944*. New York: Duell, Sloan, and Pearce.

—— (eds.). 1949. *Communications Research, 1948–1949*. New York: Harper and Brothers.

LERNER, DANIEL. 1958. *The Passing of Traditional Society: Modernizing the Middle East*. New York: Free Press.

LEVINSON, N. 1966. "Wiener's Life," *Bulletin of the American Mathematical Society* 72: 1–32.

LEWIN, KURT. 1936. *Principles of Topological Psychology*, translated by Fritz and Grace Heider. New York: McGraw-Hill.

——. 1947. "Channels of Group Life," *Human Relations* 1:145.

LEWIS, J. DAVID, and RICHARD L. SMITH. 1980. *American Sociology and Pragmatism: Mead, Chicago Sociology, and Symbolic Interaction*. Chicago: University of Chicago Press.

LIPPMANN, WALTER. 1922. *Public Opinion*. New York: Macmillan.

LOVE, GAIL, and RONALD E. RICE. 1985. "Electronic Emotion: Socioemotional Content in a Computer-Mediated Communication Network." Paper presented at the International Communication Association, Honolulu.

LUCAS, WILLIAM A. 1978. "Spartansburg, S.C.: Testing the Effectiveness of Video, Voice, and Data Feedback," *Journal of Communication* 28: 168–179.

LYND, ROBERT S. 1939. *Knowledge for What?* Princeton, N.J.: Princeton University Press.

MACCOBY, ELEANOR E., and CAROLE JACKLIN. 1974. *The Psychology of Sex Differences*. Stanford, Calif.: Stanford University Press.

MACHLUP, FRITZ. 1962. The Production and Distribution of Knowledge in the United States. Princeton, N.J.: Princeton University Press.

MARCUS, JANE ELLEN. 1985. *Diffusion of Innovations and Social Learning Theory: Adoption of the Context Text-Processing System at Stanford University*. Ph.D. thesis, Stanford, Calif.: Stanford University.

MARIEM, MICHAEL. 1984. "Some Questions for the Information Society," *Information Society* 3: 181–187.

MARROW, ALFRED J. 1969. *The Practical Theorist: The Life and Work of Kurt Lewin*. New York: Basic Books.

MARVIN, CAROLYN. 1983. "Telecommunications Policy and the Pleasure Principle," *Telecommunications Policy*, pp. 43–52.

MATTHEWS, FRED H. 1977. *Quest for an American Sociology: Robert E.*

Park and the Chicago School. Montreal: McGill-Queens University Press.

MCANANY, EMILE G. 1984. "The Diffusion of Innovation: Why Does It Endure?" *Critical Studies in Mass Communication* 1: 439–442.

MCCOMBS, MAXWELL. 1977. *Newspaper Readership and Circulation.* ANPA News Research Report 3.

MCGUIRE, WILLIAM J. 1981. "Theoretical Foundations of Campaigns," in Ronald E. Rice and William J. Paisley (eds.), *Public Communication Campaigns.* Beverly Hills, Calif.: Sage.

MCLUHAN, MARSHALL. 1965. *Understanding Media: The Extensions of Man.* New York: McGraw-Hill.

MEAD, GEORGE HERBERT. 1934. *Mind, Self and Society.* Chicago: University of Chicago Press.

MENDELSOHN, HAROLD. 1973. "Some Reasons Why Information Campaigns Can Succeed," *Public Opinion Quarterly* 39: 50–61.

MILLS, C. WRIGHT. 1959. *The Sociological Imagination.* New York: Oxford University Press.

MODY, BELLA. 1985. "First World Communication Technology in Third World Context," in Everett M. Rogers and Francis Balle (eds.), *The Media Revolution in America and Western Europe.* Norwood, N. J.: Ablex.

MOHR, LAWRENCE B. 1982. *Explaining Organizational Behavior: The Limits and Possibilities of Theory and Research.* San Francisco: Jossey-Bass.

MONGE, PETER R., and EVERETT M. ROGERS. 1985. "Investigating Process in Communication Research." Los Angeles: University of Southern California, Annenberg School of Communications. Unpublished paper.

MORENO, JACOB L. 1934. *Who Shall Survive? Foundations of Sociometry, Group Psychotherapy, and Sociodrama.* Washington, D.C., Nervous and Mental Disease Monograph 58; republished New York: Beacon House, 1953.

MOSS, MITCHELL. 1978. "Reading Pa.: Research on Community Uses," *Journal of Communication* 28: 160–167.

NAFZIGER, RALPH O. 1930. "A Reader-Interest Survey of Madison, Wisconsin," *Journalism Quarterly* 7: 128–141.

NEUMANN, JOHN VON. May 1949. "Review of *Cybernetics,*" *Physics Today,* pp. 33–34.

NOLL, A. MICHAEL. 1985. "Videotex: Anatomy of a Failure," *Information and Management* 9: 99–109.

NYQUIST, H. 1924. "Certain Factors Affecting Telegraph Speed," *Bell System Technical Journal* 13: 324.

PAISLEY, WILLIAM. 1983. "Computerizing Information: Lessons of a Videotext Trial," *Journal of Communication* 33: 153–161.

————. 1985. "Communication in the Communication Sciences," in Brenda Dervin and Melvin J. Voigt (eds.), *Progress in the Communication Sciences*, Vol. 5. Norwood, N.J.: Ablex.

————. In press. *Communication Science: The Growth of a Multidiscipline*. Norwood, N.J.: Ablex.

PARK, ROBERT E. 1922. *The Immigrant Press and Its Control*. New York: Harper.

PHILLIPS, AMY. 1982. "Computer Conferencing: Success or Failure?" *Systems, Objectives, Solutions* 2: 202–218.

POLLAK, MICHAEL. 1980. "Paul F. Lazarsfeld: A Sociointellectual Biography," *Knowledge* 2: 157–177.

POOL, ITHIEL DE SOLA (ed.). 1977. *The Social Impact of the Telephone*. Cambridge, Mass.: MIT Press.

————. 1983a. *Technologies of Freedom*. Cambridge, Mass.: Harvard University Press.

————. 1983b. "Tracking the Flow of Information," *Science* 221 (4611): 609–613.

POOL, ITHIEL DE SOLA, and HERBERT I. SCHILLER. 1981. "Perspectives on Communications Research: An Exchange," *Journal of Communication* 31: 15–23.

PORAT, MARC URI. 1978. "Global Implications of the Information Society," *Journal of Communication* 28: 70–79.

RAFAELI, SHEIZAF. 1983. "U & G of BBs: An Exploratory Study of Electronic Bulletin Boards." Stanford, Calif.: Stanford University, Institute for Communication Research. Unpublished paper.

————. 1984. "If the Computer Is the Medium, What Is the Message? Interactivity and Its Correlates." Stanford, Calif.: Stanford University, Institute for Communication Research. Unpublished paper.

————. 1985. *On Interacting with Media: Para-Social Interaction and Real Interaction*. Ph.D. thesis, Stanford, Calif.: Stanford University.

RESTON, LEO. 1969. "Harold D. Lasswell: A Memoir," in Arnold A. Rogow (ed.), *Politics, Personality, and Social Science in the Twentieth Century: Essays in Honor of Harold D. Lasswell*. Chicago: University of Chicago Press.

RICE, RONALD E. 1984. "Evaluating New Media Systems," in Jerome Johnston (ed.), *Evaluating the New Media Technologies*. San Francisco: Jossey-Bass.

RICE, RONALD E., and ASSOCIATES. 1984. *The New Media: Communication, Research, and Technology*. Beverly Hills, Calif.: Sage.

RICE, RONALD E., and CHRISTINE L. BORGMAN. 1983. "The Use of Computer-Monitored Data in Information Science and Communication Research," *Journal of the American Society for Information Science* 34: 247–256.

RICE, RONALD E., and DONALD CASE. 1983. "Electronic Message Systems in the University: A Description of Use and Utility," *Journal of Communication* 33: 131–152.

RICE, RONALD E., and EVERETT M. ROGERS. 1984. "New Methods and Data for the Study of New Media," in Ronald E. Rice et al. (eds.), *The New Media: Communication, Research, and Technology.* Beverly Hills, Calif.: Sage.

RITCHIE, DAVID. 1986. "Shannon—and Weaver? Unravelling the Paradox of Information." *Communication Research.*

ROBERTS, DONALD F., and NATHAN MACCOBY. 1985. "Effects of Mass Communication," in Gardner Lindzey and Elliot Aronson (eds.), *The Handbook of Social Psychology,* 3rd ed.

ROESSNER, J. DAVID. 1985. "Forecasting the Impact of Office Automation on Clerical Employment, 1985–2000," *Technological Forecasting and Social Change* 28.

ROGERS, EVERETT M. 1976. "Communication and Development: The Passing of the Dominant Paradigm," *Communication Research* 3: 121–133.

———. 1983. *Diffusion of Innovations.* New York: Free Press.

———. 1985a. "The Empirical and Critical Schools of Communication Research," in Everett M. Rogers and Francis Balle (eds.), *The Media Revolution in America and in Western Europe.* Norwood, N.J.: Ablex.

———. 1985b. "The Diffusion of Home Computers among Households in Silicon Valley," *Marriage and Family Review* 8: 89–101.

———. In press. "Information Technologies: How Organizations Are Changing," in Gerald M. Goldhaber (ed.), *Handbook of Organizational Communication.* Norwood, N.J.: Ablex.

ROGERS, EVERETT M., and FRANCIS BALLE (eds.). 1985. *The Media Revolution in America and in Western Europe.* Norwood, N.J.: Ablex.

ROGERS, EVERETT M., and STEVEN H. CHAFFEE. 1983. "Communication as an Academic Discipline: A Dialogue," *Journal of Communication* 33: 18–30.

ROGERS, EVERETT M., and D. LAWRENCE KINCAID. 1981. *Communication Networks: Toward a New Paradigm for Research.* New York: Free Press.

ROGERS, EVERETT M., and JUDITH K. LARSEN. 1984. *Silicon Valley Fever: Growth of High-Technology Culture.* New York: Basic Books.

ROGERS, EVERETT M., and ARNOLD PICOT. 1985. "The Impact of the New Communication Technologies," in Everett M. Rogers and Francis

Balle (eds.), *The Media Revolution in America and in Western Europe.* Norwood, N.J.: Ablex.

ROGERS, EVERETT M., and SHEIZAF RAFAELI. 1985. "Computers and Communication," *Information and Behavior* 1: 95–112.

ROGERS, EVERETT M., and J. DOUGLAS STOREY. In press. "Communication Campaigns," in Charles Berger and Steven H. Chaffee (eds.), *Handbook of Communication Science.* Beverly Hills, Calif.: Sage.

ROGERS, EVERETT M., and others. 1982. *The Diffusion of Home Computers.* Stanford, Calif.: Stanford University, Institute for Communication Research, Report.

ROGERS, EVERETT M., and others. 1985a. "The Beijing Audience Survey," *Communication Research* 12: 179–208.

ROGERS, EVERETT M., and others. 1985b. *Microcomputers in the Schools: A Case of Decentralized Diffusion.* Stanford, Calif.: Stanford University, Institute for Communication Research, Report.

ROGERS, EVERETT M., and others. 1985c. "The Diffusion of Microcomputers in California High Schools," in Milton Chen and William Paisley (eds.), *Children and Computers: Research on the Newest Medium.* Beverly Hills, Calif.: Sage.

ROWLAND, WILLARD D., JR. 1983. *The Politics of TV Violence.* Beverly Hills, Calif.: Sage.

RUBEN, BRENT D. 1984. "The Coming of the Information Age: Information, Technology, and the Study of Behavior," *Information and Behavior* 1.

RUBIN, CHARLES. August 1983. "Some People Should Be Afraid of Computers," *Personal Computing.*

RYAN, BRYCE, and NEAL C. GROSS. 1943. "The Diffusion of Hybrid Seed Corn in Two Iowa Communities," *Rural Sociology* 8: 15–24.

SCHRAMM, WILBUR. 1949. *Mass Communications.* Urbana: University of Illinois Press.

———. 1954. *The Process and Effects of Communication.* Urbana: University of Illinois Press.

———. 1955. "Information Theory and New Communication," *Journalism Quarterly* 32: 131–146.

———. 1959. "Comments on 'The State of Communication Research,'" *Public Opinion Research* 23: 6–9.

———. 1964. *Mass Media and National Development: The Role of Information in the Developing Nations.* Stanford, Calif.: Stanford University Press.

———. 1983. "The Unique Perspective of Communication: A Retrospective View," *Journal of Communication* 33: 6–17.

———. 1985. "The Beginnings of Communication Study in the United

States," in Everett M. Rogers and Francis Balle (eds.), *The Media Revolution in America and in Western Europe.* Norwood, N.J.: Ablex.

SCHRAMM, WILBUR, and others. 1961. *Television in the Lives of Our Children.* Stanford, Calif.: Stanford University Press.

SCHUDSON, MICHAEL. 1985. "Quarter Notes," *Communication Research* 12: 271–272.

SHANNON, CLAUDE E. 1948. "A Mathematical Theory of Communication," *Bell System Technical Journal* 27: 379–423, 623–656.

———. 1949. "The Mathematical Theory of Communication," in Claude E. Shannon and Warren Weaver (eds.), *The Mathematical Theory of Communication.* Urbana: University of Illinois.

SHANNON, CLAUDE E., and WARREN WEAVER. 1949. *The Mathematical Theory of Communication.* Urbana: University of Illinois Press.

SHORT, JAMES F., JR. 1971. *The Social Fabric of the Metropolis: Contributions of the Chicago School of Urban Sociology.* Chicago: University of Chicago Press.

SHORT, JOHN, and others. 1976. *The Social Psychology of Telecommunications.* New York: Wiley.

SIMMEL, GEORG. 1946. *The Web of Group-Affiliations,* translated by Reinhard Bendix. New York: Free Press.

SINGLETON, LOY. 1983. *Telecommunications in the Information Age: A Nontechnical Primer on the New Technology.* Cambridge, Mass.: Ballinger.

SLACK, JENNIFER DARYL. 1984. *Communication Technologies and Society: Conceptions of Causality and the Politics of Technological Intervention.* Norwood, N.J., Ablex.

SMITH, ANTHONY. 1980. *Goodbye Gutenberg: The Newspaper Revolution of the 1980s.* New York: Oxford University Press.

SMITH, BRUCE LANNES. 1969. "The Mystifying Intellectual History of Harold D. Lasswell," in Arnold A. Rogow (ed.), *Politics, Personality, and Social Science in the Twentieth Century: Essays in Honor of Harold D. Lasswell.* Chicago: University of Chicago Press.

SMITH, DAVID H. 1972. "Communication Research and the Idea of Process," *Speech Monographs* 39: 174–182.

STAR, SHIRLEY A., and HELEN G. HUGHES. 1950. "Report on an Educational Campaign: The Cincinnati Plan for the United Nations," *American Journal of Sociology* 55: 389–400.

STEINFIELD, CHARLES W. 1983. *Communicating via Electronic Mail: Patterns and Predictors of Use in Organizations.* Ph.D. thesis. Los Angeles: University of Southern California.

STRASSMANN, PAUL A. 1985. *Information Payoff: The Transformation of Work in the Electronic Age.* New York: Free Press.

SVENNING, LYNNE, and JOHN RUCHINSKAS. 1984. "Organizational Tele-conferencing," in Ronald E. Rice and Associates (eds.), *The New Media: Communication, Research, Technology*. Beverly Hills, Calif.: Sage.

TARDE, GABRIEL. 1903. *The Laws of Imitation*, translated by Elsie Clews Parsons. New York: Holt.

TAYLOR, WILSON L. 1953. "'Cloze Procedure': A New Tool for Measuring Readability," *Journalism Quarterly* 30: 415–433.

———. 1956. "Recent Developments in the Use of Cloze Procedure," *Journalism Quarterly* 33: 42–48.

TEHRANIAN, MAJID. 1979. "Iran: Communication, Alienation, Revolution," *Intermedia*, pp. 6–12.

THOMAS, W. I., and FLORIAN ZNANIEKI. 1927. *The Polish Peasant in Europe and America*. New York: Knopf.

TICHENOR, PHILIP J., and others. 1970. "Mass Media Flow and Differential Growth of Knowledge," *Public Opinion Quarterly* 34: 159–170.

TRIBUS, M. 1978. "Thirty Years of Information Theory," in R. D. Levine and M. Tribus (eds.), *The Maximum Entropy Formalism*. Cambridge, Mass.: MIT Press.

TURNER, RALPH H. 1967. "Introduction," in *Robert E. Park on Social Control and Collective Behavior*. Chicago: University of Chicago Press.

VENKATESH, ALLADI, and others. 1984. "Households and Technology: The Case of Home Computers—Some Conceptual and Theoretical Issues," in M. L. Roberts and L. H. Wortzel (eds.), *Marketing to the Changing Households*. Cambridge, Mass.: Ballinger.

VITALARI, NICHOLAS P., and others. 1985. "Computing in the Home: Shifts in the Time Allocation Patterns of Households," *Communications of the ACM* 28: 512–522.

WEAVER, WARREN. 1949. "Recent Contributions to the Mathematical Theory of Communication," in Claude E. Shannon and Warren Weaver (eds.), *The Mathematical Theory of Communication*. Urbana: University of Illinois Press.

WEBB, EUGENE, and others. 1981. *Nonreactive Research in the Social Sciences*. Boston: Houghton-Mifflin.

WEBER, ROBERT PHILIP. 1985. *Basic Content Analysis*. Beverly Hills, Calif.: Sage.

WHITE, DAVID MANNING. 1950. "The 'Gatekeeper': A Case Study in the Selection of News," *Journalism Quarterly* 27: 383–390.

WIENER, NORBERT. 1948. *Cybernetics, or Control and Communication in the Animal and the Machine*. Cambridge, Mass.: Technology Press, New York: Wiley.

————. 1950. *The Human Use of Human Beings: Cybernetics and Society*. New York: Houghton Mifflin.

————. 1956. *I Am a Mathematician: The Later Life of a Prodigy*. Garden City, N.Y.: Doubleday.

————. 1964. *God and Golem, Inc.: A Comment on Certain Points Where Cybernetics Impinges on Religion*. Cambridge, Mass.: MIT Press.

————. 1964b. *Ex-Prodigy*. Cambridge, Mass.: MIT Press.

WILDER, CAROL. 1979. "The Palo Alto Group: Difficulties and Directions of the Interactional View for Human Communication Research," *Human Communication Research* 5: 171–186.

Name Index

Subject Index